Gulf War Veterans

Measuring Health

Lyla M. Hernandez, Jane S. Durch, Dan G. Blazer II, and
Isabel V. Hoverman, *Editors*

Committee on Measuring the Health of Gulf War Veterans

Division of Health Promotion and Disease Prevention

INSTITUTE OF MEDICINE

NATIONAL ACADEMY PRESS
Washington, D.C.

NATIONAL ACADEMY PRESS • 2101 Constitution Avenue, N.W. • Washington, D.C. 20418

NOTICE: The project that is the subject of this report was approved by the Governing Board of the National Research Council, whose members are drawn from the councils of the National Academy of Sciences, the National Academy of Engineering, and the Institute of Medicine. The members of the committee responsible for the report were chosen for their special competences and with regard for appropriate balance.

The Institute of Medicine was chartered in 1970 by the National Academy of Sciences to enlist distinguished members of the appropriate professions in the examination of policy matters pertaining to the health of the public. In this, the Institute acts under the Academy's 1863 congressional charter responsibility to be an adviser to the federal government and its own initiative in identifying issues of medical care, research, and education. Dr. Kenneth I. Shine is president of the Institute of Medicine.

Support for this study was provided by the Departments of Defense (Contract No. DASW01-98-K-0002) and Veterans Affairs (Contract No. V101[93]P-1580). The views presented are those of the Institute of Medicine Committee on Measuring the Health of Persian Gulf Veterans and are not necessarily those of the funding organization.

International Standard Book Number 0-309-06580-1

Additional copies of this report are available for sale from:

National Academy Press
Lockbox 285
2101 Constitution Avenue, N.W.
Washington, DC 20055

Call (800) 624-6242 or (202) 334-3313 (in the Washington metropolitan area) or visit the NAP's on-line bookstore at **www.nap.edu.**

For more information about the Institute of Medicine, visit the IOM home page at **www2.nas.edu/iom.**

Printed in the United States of America

The serpent has been a symbol of long life, healing, and knowledge among almost all cultures and religions since the beginning of recorded history. The image adopted as a logotype by the Institute of Medicine is based on a relief carving from ancient Greece, now held by the Staatliche Museen in Berlin.

COMMITTEE ON MEASURING THE HEALTH OF PERSIAN GULF VETERANS

DAN BLAZER, II, M.D., Ph.D., M.P.H. (*Cochair*), Dean of Medical Education and Professor of Psychiatry and Community and Family Medicine, Office of the Dean, Duke University Medical Center, Durham, N.C.

ISABEL V. HOVERMAN, M.D. (*Cochair*), Austin Internal Medicine Associates, L.L.P., Austin, Texas, and Clinical Assistant Professor of Medicine, University of Texas Medical Branch, Galveston

MARTHA C. BEATTIE, Ph.D., Director of Research and Evaluation, Department of Alcohol and Drug Services, Santa Clara County, Calif., and Scientist, Alcohol Research Group, Berkeley, Calif.

PENNIFER ERICKSON, Ph.D., Associate Professor, Department of Health and Evaluation Services, Hershey Medical School, Pennsylvania State University, State College

NELSON GANTZ, M.D., Chairman, Department of Medicine, and Chief, Division of Infectious Diseases, Pinnacle Health Hospitals, Harrisburg, Pa., and Allegheny University of the Health Sciences

RICHARD M. GARFIELD, Dr. P.H., R.N., Bendixen Professor of Clinical International Nursing, Columbia University, New York

WILLIAM GOLDEN, M.D., Director, Division of General Internal Medicine, University of Arkansas for Medical Sciences, Little Rock

KATHLEEN N. LOHR, Ph.D., Director, Health Services and Policy Research, Research Triangle Institute, Research Triangle Park, N.C.

DAVID NERENZ, Ph.D., Director, Center for Health System Studies, Henry Ford Health System, Detroit, Mich.

DONALD L. PATRICK, Ph.D., M.S.P.H., Professor, Department of Health Services, School of Public Health and Community Medicine, University of Washington, Seattle

ROBERT O. VALDEZ, Ph.D., Professor, Health Policy and Management, University of California at Los Angeles School of Public Health

Board on Health Promotion and Disease Prevention Liaison

ROBERT B. WALLACE, M.D., Professor of Preventive and Internal Medicine, Department of Preventive Medicine, University of Iowa, Iowa City

Staff

LYLA M. HERNANDEZ, M.P.H., Study Director
JANE S. DURCH, M.A., Senior Program Officer
KELLY NORSINGLE, Project Assistant
CARA N. CHRISTIE, Project Assistant
KATHLEEN R. STRATTON, Ph.D, Director, Division of Health Promotion and Disease Prevention
DONNA D. DUNCAN, Division Assistant

Preface

Many individuals, groups, and federal agencies have a strong interest in finding answers to the numerous and complex questions regarding the health of Gulf War veterans. Various types of research and health measurement are needed to address these diverse issues. The Institute of Medicine (IOM) was asked by the Department of Veterans Affairs (VA) and the Department of Defense (DoD) to undertake a study to identify important questions concerning the health of Gulf War veterans and then to design a study to answer those questions. The committee determined that it is of fundamental importance to ask how healthy are Gulf War veterans? Are they as healthy as others? What characteristics are associated with differences between the health of Gulf War veterans and the health of others?

To address these questions, it will be necessary to measure not only the health status of those who served in the Gulf War, but also to compare Gulf War veterans with other groups. Further, one must continue to follow these groups through time to determine whether the groups differ in the way their health status is changing. As the committee began to develop a design that would address the fundamental questions identified, it realized that such a study could have important implications for understanding not only the health of Gulf War veterans, but also the health of veterans of other conflicts.

There exists a rich body of literature on the health effects of participation in specific conflicts prior to the Gulf War, including World War II and Vietnam. Research has examined the health effects of exposure to mustard gas and Agent Orange and the long-term health consequences for those who were prisoners of war. More recently, there has been renewed interest in studying poorly understood, multisymptom clusters that have been reported following every conflict since the U.S. Civil War. Similar health problems have been reported by some Gulf War veterans. Questions are beginning to emerge about whether there are health effects that are attributable to participation in military conflict in general, as well as to service in specific conflicts.

The committee recognizes that the study it is recommending will be challenging and that it will require a sustained commitment of resources by VA and DoD, and of time and cooperation by study participants. Nevertheless, we feel that these commitments are important and worthwhile if the nation is to adequately understand and respond to the health needs of Gulf War veterans.

In fact, the study designed by the committee to focus on Gulf War veterans could, with slight modification, be used to longitudinally monitor the health status of veterans of any conflict. Additionally, if a cohort were identified and the study begun immediately upon return from participation in a conflict, many of the problems we face in attempting to resolve Gulf War veterans health issues, several years removed from the end of that conflict, could be eliminated. The committee believes that such efforts would contribute greatly to our understanding of the impact of military conflict on the health of the men and women who serve in those conflicts.

In closing, we note that this committee's work complements that of several other current studies at the IOM and the National Academy of Sciences. A study on strategies to protect the health of deployed U.S. forces is scheduled for completion of the first phases in the fall of 1999 and is addressing health risk assessment and issues related to health protection, health consequences and treatment, and medical record keeping in the U.S. military services. The study of the health effects of Gulf War exposures has recently begun and will review the scientific and medical literature regarding adverse health effects associated with exposures experienced during the Gulf War. This study will include an assessment of biologic plausibility that exposures, or synergistic effects of combinations of exposures, are associated with illnesses experienced by Gulf War veterans.

We want to thank the many people, listed by name in the Acknowledgments, who contributed to this study. As cochairs of this committee, we wish to express our appreciation to the members of the committee for their insight, creativity, and hard work in developing the study approaches and methods recommended in this report. We also wish to commend Lyla Hernandez and Jane Durch for their enormous effort in producing a clearly written, well-organized report that reflects the collective thought of the committee.

Dan G. Blazer, II
Isabel V. Hoverman
Cochairs

Acknowledgments

The committee wishes to express its appreciation to the many individuals who contributed in various ways to the completion of this project. Naihua Duan, consultant to the committee, provided advice and, with Robert O. Valdez, wrote an exceptional detailed discussion of design issues to be considered in developing the Gulf War Veterans Health Study. This information appears as Appendix B in this report. Presenters at the May 1998 workshop provided an excellent overview of the health concerns of Gulf War veterans and of research efforts aimed at better understanding those concerns. The presenters were: David Cowan, Nancy Dalagar, Albert Donnay, COL John T. Graham, CAPT Gregory Gray, MAJ Charles Engel, COL Bruce Jones, Han K. Kang, Mark Meterko, Frances M. Murphy, LTC Mark Rubertone, and David A. Schwartz. Additional information was provided by Matthew Puglisi.

This report has been reviewed by individuals chosen for their diverse perspectives and technical expertise, in accordance with procedures approved by the National Research Council's Report Review Committee. The purpose of this independent review is to provide candid and critical comments that will assist the authors and the Institute of Medicine in making the published report as sound as possible and to ensure that the report meets institutional standards for objectivity, evidence, and responsiveness to the study charge. The content of the review comments and draft manuscripts remain confidential to protect the integrity of the deliberative process. The committee thanks the following individuals for their participation in the review of this report: Gerard N. Burrow, M.D., Yale University School of Medicine; Bradley N. Doebbeling, M.D., University of Iowa College of Medicine; Edward B. Perrin, Ph.D., University of Washington; Dana Gelb Safran, Sc.D., New England Medical Center; and Jonathan M. Samet, M.D., Johns Hopkins University.

Although the individuals listed above have provided many constructive comments and suggestions, responsibility for the final content of this report rests solely with the authoring committee and the IOM.

Contents

EXECUTIVE SUMMARY 1
Committee Charge, 3
Conclusions and Recommendations, 6
Summary, 10

1 **INTRODUCTION** 12
VA and DoD Programs for Gulf War Veterans, 13
Gulf War Reports and Evaluations: A Brief Summary, 17
Measuring the Health of Gulf War Veterans, 20
Defining Gulf War Veterans, 22
Structure of Report, 23

2 **STUDIES OF THE HEALTH OF GULF WAR VETERANS** 24
Introduction, 24
Specific Studies of Gulf War Veterans, 25
Limitations of Previous Studies, 36
Remainder of this Report, 37

3 **MEASURING HEALTH** 38
The Evolving Definition of Health, 38
Core Concepts of Health, 40
Correlates of Health, 45
Measuring Health-Related Quality of Life, 48
Summary, 54

4 **THE GULF WAR VETERANS HEALTH RESEARCH PORTFOLIO** 55
Research Portfolio to Guide Studies of the Health of Gulf War Veterans, 55
Summary, 61

5 GULF WAR VETERANS HEALTH STUDY 63
Study Questions and Design, 64
Sampling, 67
Scheduling, 70
Mode of Survey, 71
Improving Response Rates, 71
Pilot Study, 75
Data Collection Instruments, 76
Cost, 78
Ethical Considerations, 79
Independent Advisory Board, 80
Summary, 81

6 CONCLUSION 83

REFERENCES 87

APPENDIX A: COALITION FORCES AND FORCE STRENGTH 95

APPENDIX B: DESIGN ISSUES IN THE GULF WAR VETERANS HEALTH STUDY 96

Executive Summary

The Gulf War was short in duration but the consequences linger 8 years after the fighting ceased. Some veterans of that conflict report debilitating health problems that they believe are connected to service in the Gulf.* Once healthy and fit soldiers report they are no longer able to engage in normal daily activities, much less the rigorous tasks they completed in the military. Symptoms commonly described include fatigue, memory loss, severe headaches, muscle and joint pain, and rashes (Fukuda et al., 1998; Iowa Persian Gulf Study Group, 1997). These veterans want to know why they are ill, what can be done to make them better, and whether the government is doing all it can to help them. They have taken their case to the media, to Congress, to the Department of Veterans Affairs (VA), and to the Department of Defense (DoD).

With concern about the veterans' reports escalating, numerous activities were launched to investigate veterans' health concerns. Various aspects of the problem have been studied by a Presidential Advisory Committee (PAC), the General Accounting Office (GAO), a special investigation unit of the Committee on Veterans Affairs of the U.S. Senate, the Centers for Disease Control and Prevention (CDC), the Institute of Medicine (IOM), and independent researchers. The federal government has spent more than $230 million to fund research efforts and diagnostic programs to answer the many questions raised.

Research studies have compared Gulf War veterans to other contemporary military veterans to determine whether they have higher hospitalization rates

---

*For purposes of this report, the term "veteran" refers to any person who served on active duty or in the reserves or National Guard during the period of the Gulf War or other specified engagements. A Gulf War veteran is defined as any person who served on active duty in the Gulf War theater of operations between August 2, 1990, and June 13, 1991. Thus, the "veteran" population may include persons who remain on active duty or continue to serve in the reserves or National Guard.

(Gray et al., 1996; Knoke and Gray, 1998), a greater incidence of reproductive problems (Araneta et al., 1997; Cowan et al., 1997), or higher mortality rates (Kang et al., 1996). One study has compared the health of Iowa Gulf War veterans to that of Iowa veterans who were not deployed to the war (Iowa Persian Gulf Study Group, 1997). Reviews of scientific literature and new research have been conducted to try to determine whether any veterans' Gulf War exposures could be responsible for their symptoms, and various approaches to treatment of veterans' problems have been tried.

The findings of these studies, evidence from the many efforts to evaluate Gulf War veterans' health conducted by the aforementioned bodies, and additional independent research support several conclusions:

- No single diagnosable illness or set of symptoms *with a known etiology* characterizes either Gulf War veterans in general or a subset of veterans who are experiencing some kind of health problem.
- VA and DoD data systems do not demonstrate a higher incidence of hospitalizations or deaths among Gulf War veterans than among other veterans. The incidence of birth defects in children and the incidence of health problems among spouses are not higher for the Gulf War cohort than for other veteran cohorts.
- There *does* seem to be a higher prevalence of some symptoms among veterans who served in the Gulf War as compared to nondeployed veterans. The primary symptoms include fatigue, difficulty concentrating, memory loss, skin rash, headache, and muscle and joint pain.
- Many Gulf War veterans receive no diagnosis that explains their symptoms. Many of these complaints produce no observably physiological indicators, and must be measured by self-reports of those experiencing them. For some veterans, the symptoms are severe enough to be disabling; others experience milder symptoms that still allow some level of normal daily activity, while others report no problems at all.
- Several possible explanations for the symptoms experienced by Gulf War veterans have been suggested. They include exposure to vaccines, toxic chemicals, chemical and biological warfare agents, and depleted uranium, as well as stress associated with either exposure to battlefield stimuli or rapid deployment with associated uncertainty about time and circumstances of return home. Veterans' symptoms have not been found to be correlated with exposure to any particular physical or psychological stimulus. Although future epidemiological studies may show such an association, the extant literature is limited and results are inconclusive.
- No strong evidence exists for any effective treatment of these symptoms *as a single disease entity*. There is little published evidence of successful treatment of symptom clusters.

Many areas of uncertainty remain. For example, no one has yet determined the extent of the problem, that is, the number of veterans who have symptoms or illnesses that they attribute to service in the Gulf War. Also, no one yet knows whether the health status of the Gulf War veteran population is better than, worse than, or the same as that of veterans who were not deployed to the Gulf War, although some studies have found higher levels of reported symptoms among Gulf War veterans. There has been no systematic evaluation of whether the health status of these veterans is changing and, if so, how.

COMMITTEE CHARGE

In December 1997, the VA and the DoD asked the IOM to convene a group of experts to consider the numerous questions regarding the health of Gulf War veterans and then to determine how best to address the issues of measuring and monitoring the health of these veterans. The charge to the IOM was to: "(1) identify relevant questions regarding the evaluation of the health status of active-duty troops and veterans deployed to the Gulf War; (2) identify issues to be addressed in the development of study designs and methods that would be used to answer such questions; and (3) develop a research design(s) and methods that could be used to address such questions."

The IOM convened the Committee on Measuring the Health of Gulf War Veterans, which is composed of experts in outcomes analysis, study design, research methods, statistics, epidemiology, health status measurement, military health databases, clinical medicine, and Gulf War veterans' health. Between May 1998 and April 1999, the committee met five times. In addition, a workshop was held in May 1998 to obtain background information on the health concerns of Gulf War veterans and an overview of relevant research. During subsequent meetings the Committee reviewed and analyzed additional information on: symptoms, complaints, and diagnoses of veterans; completed population-based and sample-survey research on the health of Gulf War veterans from the United States, Canada, and the United Kingdom; VA and DoD health databases; the reports of the PAC, the GAO, and other IOM committees; and books and articles describing and evaluating approaches to and instruments for measuring health status.

The first component of the study charge directed the committee to identify questions important in evaluating the health and well being of active-duty troops and veterans who were deployed to the Gulf War. Through a review of statements and presentations by major interested groups, the committee identified questions that appear critical to those groups (see Table 1). Some of these questions can be addressed by research, others are in the realm of policy.

Furthermore, many individuals and groups are now beginning to ask if these questions apply only to the health of Gulf War veterans or if they also apply to the health of veterans of any conflict.

TABLE 1 Questions, by Group Asking, About the Health of Gulf War Veterans

Veterans[a]	• How many Gulf War veterans are ill and why? • Are we getting the care we need? • Will we get better or will we get worse? • Has all the information on Gulf War exposures and health problems been made public? • How can we get the government to listen to us?
Department of Veterans Affairs[b]	• Longitudinal follow-up of health status of Gulf War veterans is challenging, especially for ill-defined or undiagnosed conditions. What methodological questions should be considered regarding acquiring and analyzing longitudinal information? • What scientific studies should be conducted to resolve the areas of continued scientific uncertainty related to health outcomes and treatment efficacy in Gulf War veterans? • What would be the best approach to answering questions regarding the health of Gulf War veterans? Would it be a national database? Well-designed research studies? A longitudinal study? A study similar to the Ranch Hands study of Vietnam veterans? • How can we obtain a better understanding of treatment for chronic fatigue syndrome, fibromyalgia, and PTSD?
Department of Defense[b]	• Are treatments improving the health of Gulf War veterans and, if not, are veterans being medically followed? • Are clinical trials needed to answer questions about Gulf War veterans' health and, if so, for what entities? • Should DoD and VA examine clusters of illnesses and clusters of symptoms and their treatment and then measure outcomes? • Are there known treatment interventions that DoD is not using that would be more successful with some of these difficult conditions?

Continued

TABLE 1 *Continued*

Department of Defense (*continued*)	• Do we need a longitudinal cohort study to answer questions about the health of Gulf War veterans? • Should questions regarding mental health be considered? • Do these questions apply to all deployments and not just the Gulf War?
Congress[c]	• How many veterans are ill? Is the number increasing? • What is happening to the sick veterans? Are they getting any better or are they getting worse? • Why don't we know what is wrong with the ill Gulf War veterans? • Are treatment trials needed to determine what will help these veterans? • What are VA and DoD doing to help the veterans? • Why did it take so long for anyone in DoD or VA to recognize this problem?
General Accounting Office[d]	• Are ill Gulf War veterans in better or worse physical health than when they were first examined? What is their clinical progress over time? • Are Gulf War veterans receiving appropriate, effective, high-quality health care? • What plans do DoD and VA have to provide data on effectiveness of treatments received by Gulf War veterans? • What plans do DoD and VA have to collect longitudinal information on the health of veterans who report diagnosed and undiagnosed illnesses after the war? • How can we know if Gulf War veterans are ill due to something that happened in the Gulf War, since adequate exposure data or adequate data on their health at that time are not available?

[a]From testimony presented by veterans and veterans organizations during congressional hearings.
[b]Questions raised by VA and DoD during formal presentation of the study charge to the IOM committee.
[c]Questions asked by members of Congress during congressional hearings.
[d]GAO, 1997, 1998.

CONCLUSIONS AND RECOMMENDATIONS

The committee concluded that a single study cannot satisfy all information needs concerning the health of Gulf War veterans. Because these questions are diverse and require the application of various types of research and health measurement to address them thoroughly, the committee has developed and recommends a "portfolio" of research activities that includes population studies, health services research, and clinical and biomedical investigations.

Of fundamental importance is the need to know how many Gulf War veterans are suffering from health problems that affect their ability to function; whether the prevalence of such problems among Gulf War veterans is consistent with their prevalence among the general public or among other veterans groups; and whether the health of veterans is getting better, staying the same, or deteriorating with time.

Because these fundamental questions address both the health of Gulf War veterans at specific time points and changes over time, **the committee recommends that a prospective cohort study of the population of Gulf War veterans be conducted. Such a study should include appropriate comparison groups.**

Additionally, **the committee recommends that the prospective cohort study investigate the following four questions:**

1. How healthy are Gulf War veterans?

2. In what ways does the health of Gulf War veterans change over time?

3. Now and in the future, how does the health of Gulf War veterans compare with that of

- **the general population;**
- **persons in the military at the time of the Gulf War but not deployed;**
- **persons in the military at the time of the Gulf War who were deployed to nonconflict areas; and**
- **persons in the military deployed to other conflicts, such as Bosnia, Somalia?**

4. What individual and environmental characteristics are associated with observed differences in health between Gulf War veterans and comparison groups?

Key comparison groups must be included in the study design to provide a basis for drawing conclusions about reasons for levels and trends in the health status of Gulf War veterans. Comparisons with the general population provide a basis for ascertaining whether the health experience of Gulf War veterans simply reflects levels and trends in the general population, or is different because of some aspect of military service. Additionally, a general population comparison group will provide a basis for distinguishing changes in health of Gulf War veterans that are attributable to factors such as economic conditions experienced by veterans since the War versus participation in the War. For example, decline in health may be attributable to economic recession, as was found for the general population in an analysis of the Health and Activity Limitation Index over the 1984–1994 interval (Erickson, in press). Comparisons with those in the military at the time of the Gulf War but not deployed provide a basis for ascertaining whether selection for deployment is associated with differences in health levels and trends.

Comparisons with a sample of individuals who were in the military and who were deployed to theater but were deployed to a "safe" area provide a basis for ascertaining whether selection for war-theater deployment had health consequences or whether health levels and trends resulted simply from selection for deployment to the Gulf region. Finally, comparisons with a sample of veterans of other conflicts (e.g., Bosnia and Somalia) provide a basis for ascertaining whether health levels and trends of Gulf War veterans are a consequence of serving in any conflict or the result of unique aspects of the Gulf War situation.

The committee recognizes that many completed studies have made important contributions to our understanding of the issues and problems affecting the health of Gulf War veterans and that other valuable studies are currently underway or will be undertaken in the coming years. Various agenda-setting bodies are directing the flow of resources to these investigations. The committee believes, however, that the contributions of future individual studies will be enhanced if a mechanism exists for linking these studies through the collection of a core set of key data elements, thereby allowing comparisons across all research undertaken. Linking studies in this manner is an essential feature of the committee's research portfolio concept.

The committee believes that this portfolio approach, which encompasses various study designs and related methods, will, if implemented, lead to a greater understanding of the longer-term health effects of service in the Gulf War. Therefore, **the committee recommends that multiple studies be initiated through a research portfolio with three components: population studies, health services research studies, and biomedical and clinical investigations.**

The committee recommends that a core set of data on health be collected in all studies and include measures of:

- **death and duration of life,**
- **impairment,**
- **functional status,**
- **health perceptions, and**
- **opportunity [the capacity for health, the ability to withstand stress, and physiological reserve].**

The committee recommends that a core set of data on the correlates of health be collected in all studies. These data should include measures of individual and environmental characteristics that are associated with differences in health. Individual characteristics of interest include:

- **biology and life course,**
- **lifestyle and health behavior,**
- **illness behavior,**
- **personality and motivation, and**
- **values and preferences.**

Environmental characteristics of interest include:

- **social and cultural ,**
- **economic and political,**
- **physical and geographic, and**
- **health and social care.**

The committee further recommends that the prospective cohort study of Gulf War veterans (and appropriate comparison groups) serve as the foundation for the entire portfolio of activities.

The committee has not attempted to develop detailed design specifications for such a study, but it has identified key methodological considerations. Specifically, **the committee recommends the prospective cohort study incorporate the following features:**

- **multiple cohorts, one for each group of interest;**
- **multistage sampling with initial cluster sampling followed by stratified random sampling within clusters; and**
- **random and representative selection of participants within clusters; hypothesis-driven oversampling of specific population subgroups; and multiple modes of interviewing, including telephone and in-person interviewing.**

Although the committee is persuaded that a prospective cohort study is a necessary and appropriate method for monitoring the health of Gulf War veterans, it recognizes that such a study requires a major commitment of resources. Therefore, **the committee recommends a pilot study be conducted to determine the feasibility and cost of the prospective cohort study. The pilot study should include an assessment of the following points:**

- **for each of the five cohorts, identification of the universe from which the sample is to be drawn, especially the Gulf War veteran sample;**
- **willingness of members of each cohort to participate in the baseline study;**

- **modes of data collection; and**
- **use of incentives to maximize response rates.**

An independent advisory board should be established to ensure high-quality research throughout the program and to set policies for and monitor the progress of the long-term survey and research portfolio of studies of the health of Gulf War veterans. This advisory board should be independent in order to ensure its scientific integrity and the public perception of validity of research results.

The committee sees several benefits of such an advisory board. First, it would provide a means for engaging a broad range of expertise in the oversight of this major, and complicated, effort to monitor and improve the health of veterans of military conflict. Second, its agenda can be quite broad and encompass more than might be accomplished by any single federal department. Third, it would provide a visible mechanism for public accountability. Finally, such an advisory board can command national attention when it speaks or acts; it is thus in a position to call for direct, immediate, and meaningful action on the conclusions and implications of critical findings.

Specific functions of the advisory board should include a review of the scientific and methodological merit of proposed and ongoing studies in the research portfolio. This review would take into account not only the research activities being supported or carried out within the structure proposed in this report, but also changes in various other programs within the federal government and the private sector.

Specifically, **the committee recommends that an independent advisory board oversee the conduct of the prospective cohort study. The advisory board should**

- **be an independent, scientific, and policy-oriented body composed of experts in clinical medicine, epidemiology, health status and health outcomes assessment; veterans' health issues; health services research; social, behavioral, physical, and biomedical sciences; survey research; statistics; national health databases; health policy; and members of the public who represent Gulf War veterans.**
- **review, in a timely fashion, requests for proposals developed by the funding agencies to conduct the prospective cohort study recommended by the committee.**
- **evaluate the methodological design of the cohort study.**
- **set the minimum requirements for policies on:**

  - **methods for locating and retaining study participants,**
  - **informed consent,**
  - **respondent burden,**
  - **confidentiality and security of data,**
  - **use of incentives, and**

— responsibility for reporting identified individual or public health threats.

- **evaluate the success of the prospective cohort study at the end of the 10-year study period.**
- **submit a report to Congress every 2 years.**

SUMMARY

As this report is being prepared, 8 years have elapsed since the last U.S. troops returned from the Gulf War. During that time enormous effort has been expended in attempts to solve the puzzle of the effects of the Gulf War on the health of those deployed to fight in that war. Veterans have lobbied extensively to ensure that their concerns are heard and problems are addressed. Numerous investigations by Congress, the GAO, the PAC, and the IOM have attempted to tease out factors contributing to those problems. VA and DoD have established examination programs focused on diagnosing and treating Gulf War veterans' complaints. The Department of Health and Human Services (HHS), VA, and DoD have funded more than 120 research projects aimed at various aspects of the problem. Independent researchers have engaged in additional research focused on the health of Gulf War veterans.

VA has undertaken the enormous task of coordinating research efforts through the Research Working Group of the Persian Gulf Veterans Coordinating Board, chaired by the secretaries of Defense, HHS, and VA. Yearly reports of activity have been submitted to Congress. Much has been learned, but much remains to be accomplished.

The Committee on Measuring the Health of Gulf War Veterans believes that the recommendations in this report will contribute answers to many of the remaining questions. We must learn the extent of the health problems experienced by Gulf War veterans, both those using the DoD and VA health systems and those seeking health care in the private sector. We must also ascertain how and in what ways the health problems of Gulf War veterans differ from those of the general public and other veterans groups. We must determine how the health status of all these groups changes through time, thereby enabling us to understand whether and to what extent Gulf War veterans differ from other groups. The prospective cohort study designed by the committee will provide the basis for answering these fundamental questions.

Additionally, given the enormous amount of time, effort, and resources devoted to numerous studies of the health of Gulf War veterans, it is important that a mechanism be implemented to allow comparisons across these studies on key health correlates and health outcomes. The research portfolio recommended by the committee is designed to accomplish this.

Finally, to assure the public, the veterans, Congress, the scientific community, and others that all efforts to resolve these issues are being conducted with the greatest scientific integrity and public accountability, the committee believes it is necessary to establish an independent advisory board to oversee the implementation of the prospective cohort study and related research portfolio, with periodic reports on these efforts to Congress.

1

Introduction

On August 2, 1990, Iraq invaded the independent nation of Kuwait. Within 5 days the United States began to deploy troops to the region. Ultimately, in response to United Nations Resolution 678, a coalition of 41 countries mobilized a force of almost 1 million soldiers, 700,000 of whom were U.S. troops. (See Appendix A for a list of participating countries and numbers of troops.)

From August 1990 through early January 1991, troops settled into position and prepared for war. On January 16, 1991, intense air attacks against the Iraqi forces were begun, and on February 24 a ground attack was launched. Within 4 days Iraqi resistance crumbled. Following the fighting, the number of troops in the area declined rapidly. By June 13, 1991, the last U.S. troops who participated in the ground war returned home.

The demographic characteristics of the U.S. troops deployed to the Gulf War differed from those involved in previous military engagements. Overall, they were older, a large proportion (about 17%) were from National Guard and Reserve units, and almost 7% of the total forces were women.

U.S. casualties were low during the Gulf War. There were 148 combat deaths, with an additional 145 deaths due to disease or injury. Deployed personnel were, however, exposed to a number of stressors. The term stressor generally refers to the external circumstances that challenge or obstruct an individual (IOM, 1997a). Stressors to which those deployed to the Gulf War may have been exposed are listed in Table 1-1.

Following the war, most veterans returned home and resumed their normal activities. Within a relatively short time, however, some began to report health problems they believed were connected to their service in the Gulf. Commonly reported problems include fatigue, moodiness, cognitive problems, muscle and joint pain, shortness of breath, and rashes (Fukuda et al., 1998; Iowa Persian Gulf Study Group, 1997).

TABLE 1-1 Stressors of the Gulf War

Chemical	Environmental	Combat Related
Oil fire smoke	Sand	Rapid mobilization leading to unexpected disruption of lives, particularly for Reserve and Guard units
Diesel and jet fuel	Fleas and other insects	Waiting for combat to begin
Solvents and other petrochemicals	Extreme heat	Potential cumulative effect of repeated deployments to conflict
Insect repellents	Relatively primitive living conditions	Rapid demobilization, particularly for Reserve and Guard units
CARC paint	Unfamiliar character of region	SCUD missile attacks
Depleted uranium	Prohibition against interaction with indigenous population	Multiple chemical alarms
Anthrax and botulinum vaccines	Exposure to dead and mutilated bodies	
Pyridostigmine bromide pills	Exposure to dead animals	

SOURCE: IOM, 1996b, 1998.

There have been a variety of responses to these reports. The Department of Veterans Affairs (VA) and the Department of Defense (DoD) developed programs to examine Gulf War veterans and diagnose and treat their illnesses. Researchers and policymakers also responded with studies to characterize the veterans' health problems, explore potential causes of those problems, and assess the adequacy of the response by VA and DoD.

VA AND DoD PROGRAMS FOR GULF WAR VETERANS

VA Persian Gulf Registry and Uniform Case Assessment Protocol

In 1992, the VA developed and implemented the Persian Gulf Registry. The original purposes of the registry were to

• create a database containing medical and other data on Gulf War veterans that would assist in addressing questions about possible future effects of exposures to air pollutants and other environmental agents; and

- serve as the basis for future medical surveillance (VA, 1995).

Exposures, particularly those associated with the oil well fires, were included as part of the veterans' history taking. As time passed, it became apparent that several exposure issues and a host of symptoms needed further investigation.

The VA diagnostic program for Gulf War veterans is divided into two phases, the Registry Exam and the Uniform Case Assessment Protocol (UCAP). During the Registry Exam, a complete medical history is taken; time of onset of symptoms or condition, intensity, the degree of physical incapacitation and details of any treatment received are recorded; and basic laboratory tests are administered.

UCAP provides for the additional examination and testing given to those veterans who, after completing the registry evaluation, are found to have a disability but no clearly defined diagnosis that explains their health problems. Symptom-specific supplemental baseline laboratory tests and consultations are ordered. If a diagnosis is not made after completing the UCAP investigation of the veteran's symptoms and conditions, the veteran may be referred to one of four Gulf Referral Centers. These centers offer inpatient stays during which observation, multidisciplinary consultation, documentation of lengthy occupational and exposure histories, and serial physical examinations are conducted.

DoD Comprehensive Clinical Evaluation Program

In 1994, the DoD implemented a clinical diagnostic program similar to that of the VA, called the Comprehensive Clinical Evaluation Program (CCEP). This program was intended to provide a thorough, systematic clinical evaluation for the diagnosis of health problems of Gulf War veterans. Specifically, the CCEP was designed to (1) assess possible Gulf War-related conditions; (2) streamline patient access to medical care; (3) make clinical diagnoses in order to treat patients; (4) provide a standardized, staged evaluation and treatment program; and (5) strengthen the coordination between DoD and the VA in the provision of health care (IOM, 1997a).

The CCEP is a two-phase process as is the VA program. It consists of a medical history, physical examinations, and laboratory tests. All participants in the CCEP are evaluated by a primary care physician at their local medical treatment facility and receive specialty consultations, if the primary care physician deems them to be appropriate. Evaluation at this phase includes a survey for nonspecific patient symptoms, including fatigue, joint pain, diarrhea, difficulty concentrating, memory and sleep disturbances, and rashes.

Primary care physicians may refer patients to Phase II for further specialty consultations if they determine that such referrals are clinically indicated. These Phase II evaluations are conducted at a regional medical center and consist of targeted, symptom-specific examinations, lab tests, and consultations.

In March 1995, DoD established the Specialized Care Center at Walter Reed Army Medical Center to provide additional evaluation, treatment, and rehabilitation for patients suffering from chronic debilitating symptoms. The Specialized Care Program consists of an intensive 3-week evaluation and treatment protocol designed to improve the health status of participants. As of March 1999, the program had served 200 patients.

Participation in the VA and DoD Programs

Participation in the VA and DoD special diagnostic programs for Gulf War veterans is voluntary. By March 1999, almost 125,000 Gulf War veterans had participated in one of these programs, 72,000 in the VA Registry and 52,000 in the CCEP. Table 1-2 displays some of the basic demographic and military service characteristics of the first 83,000 participants as of late 1997, the latest date for which complete information is available for both. As can be seen from these data, the demographic and military characteristics of those participating in the registries differ from the distributions of those characteristics in all Gulf War veterans.

TABLE 1-2 Demographic and Military Service Characteristics of All Gulf War Veterans and Veterans Participating in the VA PGR/UCAP or the DoD CCEP

Demographic or Military Service Characteristics	% All Gulf War Veterans[a] (n = 696,530)	% PGR/ UCAP[b] (n = 57,253)	% CCEP[c] (n = 27,747)	% PGR or CCEP[d] (n = 83,197)
Age Group (yrs.) (1991)				
<25	42.0	37.0	24.2	33.1
25–34	39.7	34.6	48.9	39.4
35–44	15.5	21.5	23.7	21.9
45–54	2.6	6.2	2.9	5.0
55–64	0.2	0.7	0.2	0.6
≥65	0.0	0.0	0.0	0.0
Sex[e]				
Male	89.1	89.8	89.7	89.9
Female	6.9	10.2	10.3	10.1
Unknown	4.0	0.0	0.0	0.0
Race				
White	65.3	64.5	55.4	61.5
Black	21.8	23.6	32.9	26.6
Hispanic	4.8	5.5	5.1	5.4
American Indian	0.6	0.8	0.6	0.7
Asian	2.2	1.1	1.5	1.2
Other	1.3	1.3	2.3	1.6
Unknown	4.1	3.2	2.3	3.0

Continued

TABLE 1-2 *Continued*

Demographic or Military Service Characteristics	% All Gulf War Veterans[a] (*n* = 696,530)	% PGR/ UCAP[b] (*n* = 57,253)	% CCEP[c] (*n* = 27,747)	% PGR or CCEP[d] (*n* = 83,197)
Marital Status				
Married	48.0	49.8	66.4	54.9
Single	45.1	42.6	26.8	37.7
No longer married	2.8	4.4	4.5	4.4
Unknown	4.1	3.2	2.3	2.9
Highest Level of Education				
Elementary school	0.5	1.4	0.3	1.0
High school	1.8	1.3	0.8	1.1
High school diploma	73.7	74.9	73.7	74.7
Some college	2.5	2.8	5.0	3.5
Bachelor's degree	10.0	10.2	8.3	9.5
Master's degree	2.2	1.4	2.2	1.6
Post-master's degree	4.1	4.0	6.1	4.7
Other/unknown	5.2	4.1	3.5	3.9
Branch				
Army	50.4	73.3	84.5	76.8
Air Force	11.9	6.4	6.8	6.5
Marine Corps	14.9	13.0	5.0	10.5
Navy	22.7	7.1	3.7	6.0
Coast Guard	0.1	0.2	0.1	0.2
Pay Grade				
Enlisted	89.3	93.2	89.9	92.2
Officer	9.5	5.6	7.7	6.3
Warrant	1.2	1.1	2.4	1.5
Military Component				
Active duty	83.9	60.7	91.1	70.9
Reserve and guard	16.1	39.3	8.9	29.1

[a]Gulf War veterans include military personnel deployed to the Gulf between August 1, 1990, and July 31, 1991.
[b]VA PGR/UCAP participants from inception to November 27, 1997.
[c]DoD CCEP participants from inception to December 10, 1997.
[d]Unique veterans from both registries; 1,803 individuals are enrolled in both registries.
[e]If sex was unknown in the Defense Manpower Data Center roster, sex as recorded in the registry was used to tabulate the data.

SOURCE: Department of Veterans Affairs, 1998b.

GULF WAR REPORTS AND EVALUATIONS: A BRIEF SUMMARY

As concern increased about Gulf War veterans' health problems and their causes, the media began to talk about "Gulf War Syndrome," an illness they described as characterized by the long and growing list of symptoms being experienced by Gulf War veterans. Congress enacted legislation aimed at providing medical care to veterans experiencing problems and called for investigations into the causes.

U.S. General Accounting Office

The U.S. General Accounting Office (GAO) conducted several evaluations of Gulf War veterans' health problems and the VA and DoD responses. GAO criticized the Army's preparation for and response to depleted uranium (DU) exposure during the Gulf War (GAO, 1993). A report on the health concerns of Gulf War veterans from the 123rd Army Reserve Command headquartered in Indianapolis, Indiana (GAO, 1995), found that most of these veterans reported having health problems that limited their physical and social activities to some extent and that veterans believed these problems were caused by their service in the Gulf. The veterans were either dissatisfied with the medical services received from DoD and VA or were unaware such services were available. A June 1997 report was extremely critical of the DoD and VA efforts to monitor the clinical progress of Gulf War veterans (GAO, 1997). In an investigation of tumors in Gulf War veterans, GAO reported that incidence could not be reliably determined from available data (GAO, 1998a). A subsequent report observed that "[w]hile the number of Gulf War veterans who participated in the military operations known as Desert Shield and Desert Storm is well established at almost 700,000, the number who actually suffer, or believe they suffer, from illnesses related to their Gulf War service remains uncertain 7 years after the war" (GAO, 1998b:2). This report recommended that VA provide a case management approach to the care of Gulf War veterans, and that VA work to fully and uniformly implement these systems in their facilities.

Presidential Advisory Committee on Gulf War Veterans' Illnesses

In May 1995, President Clinton established the Presidential Advisory Committee on Gulf War Veterans' Illnesses (PAC) to conduct an independent and comprehensive review of health concerns related to Gulf War service. This 12-member panel reviewed research, coordination efforts, medical treatment, outreach, reviews conducted by other governmental and nongovernmental bodies, exposures and health effects, and the possibility that chemical and biological weapons were used in the Gulf.

The PAC deliberations resulted in six key conclusions.

1. Although the government had been somewhat slow to act at the end of the Gulf War, it was now providing appropriate medical care to Gulf War veterans.
2. The government's research portfolio was appropriately weighted toward epidemiological studies and studies on stress-related disorders.
3. DoD investigations into possible chemical and biological warfare agent exposures had produced an atmosphere of mistrust surrounding every aspect of Gulf War veterans' illnesses, and the government had lost credibility with the public.
4. Many veterans have illnesses that are likely connected to their service in the Gulf.
5. There is no evidence of a causal link between reported symptoms and illnesses and specific exposures.
6. Stress is likely to be an important contributing factor in these illnesses (Presidential Advisory Committee, 1996a,b, 1997).

Institute of Medicine

The Institute of Medicine (IOM) undertook several activities focusing on the potential health implications of deployment in the Gulf War and on the responses by the DoD and the VA to address veterans' health concerns. The IOM Medical Follow-up Agency examined the health consequences of service in the Gulf and developed recommendations for research and information systems. The first report of this group (IOM, 1995:8) recommended that "the VA Persian Gulf Health Registry should be limited and specific to gathering information to determine the types of conditions reported. There should be efforts to implement quality control and standardization of data collected by the registry." The report also recommended improved outreach to inform veterans about the availability of the registry. A second report focused on findings and recommendations concerning research and information systems needed to assess the health consequences of service during the Gulf War (IOM, 1996b).

An evaluation of the adequacy of the DoD CCEP concluded that, even though the CCEP was a comprehensive effort to address the clinical needs of those who had served in the Gulf War, specific changes in the protocol would help increase its diagnostic yield (IOM, 1996a). The study also concluded that the CCEP was not appropriate as a research tool but that the results could and should be used to educate Gulf War veterans and the physicians caring for them, to improve the medical protocol itself, and to evaluate patient outcomes.

IOM continued its evaluation of the CCEP, focusing attention specifically on difficult-to-diagnose problems and ill-defined conditions, the diagnosis and treatment of stress and psychiatric conditions, and the assessment of health problems of those who may have been exposed to low levels of nerve agents. The report addressing the adequacy of the CCEP relative to nerve agents concluded that the CCEP provided an appropriate screening approach to the diagnosis of neurological diseases and conditions but recommended certain refinements to enhance the program (IOM, 1997b).

In addressing the issues of medically unexplained conditions, and stress and psychiatric disorders, the IOM (1997a) emphasized the need to treat veterans' symptoms whether or not there had been a diagnosis; the need to provide increased screening for depression, traumatic exposure, and substance abuse; the importance of conducting an evaluation across facilities to determine consistency in terms of examination and patterns of referral; and the need for greater coordination between the DoD and the VA, particularly relating to the ongoing treatment of patients.

A separate IOM committee evaluated the adequacy of the VA medical program for Gulf War veterans (IOM, 1998). Its report complimented the VA for its overall provider education and outreach efforts. Recommendations called for the development of clinical practice guidelines for the difficult-to-diagnose or unexplained symptom constellations, and for the establishment of a system of feedback and continuous quality improvement to monitor the care received by Gulf War veterans.

Other Investigations

In April 1997, the U.S. Senate Committee on Veterans' Affairs created an expert bipartisan special investigation unit (SIU) to undertake a comprehensive and detailed review of what may have caused the illnesses of Gulf War veterans. This unit also investigated what should be done to treat these veterans and how to avoid such uncertainty in future situations. Its report found that "while there does not appear to be any single 'Gulf War syndrome,' there is a constellation of symptoms and illnesses whose cause or causes eludes explanation at this time" (U.S. Senate Committee on Veterans' Affairs, 1998:3). Further, there is a great need to monitor Gulf War veterans to determine whether their health is getting better or worse and to define the long-term health effects they may experience.

The Executive Office of the President (EOP, 1998) issued a report making several recommendations aimed at improving the federal government's responses to the health needs of its military, veterans and their families. When health problems are identified following a military deployment, the report states, there must be plans in place to improve and facilitate cooperation and coordination among DoD, VA, and HHS. The report recommended (1) creating a Military and Veterans Health Coordinating Board, (2) developing an Information Management/Information Technology Task Force, and (3) implementing strategies aimed at deployment health, record keeping, health risk communication, and research (EOP, 1998).

MEASURING THE HEALTH OF GULF WAR VETERANS

Committee Charge and Activities

Because of continuing questions and concerns about the health of Gulf War veterans, VA and DoD asked IOM in December 1997 to convene a group of experts to consider the numerous questions and determine how best to address the issues of measuring and monitoring the health of Gulf War veterans. The charge to IOM was to (1) identify relevant questions regarding the evaluation of the health status of active-duty troops and veterans deployed to the Gulf War; (2) identify issues to be addressed in the development of study designs and methods that would be used to answer such questions; and (3) develop a research design(s) and methods that could be used to address such questions.

In response to the request from DoD and VA, IOM convened the Committee on Measuring the Health of Gulf War Veterans. The committee is composed of experts in outcomes analysis, study design, research methods, statistics, epidemiology, health status measurement, military health databases, clinical medicine, and Gulf War veterans' health. As a starting point for its deliberations, the committee held a workshop on May 7, 1998, to obtain background information on the health concerns of Gulf War veterans and an overview of relevant research. During subsequent meetings the committee reviewed and analyzed additional information on the following topics: symptoms, complaints, and diagnoses of veterans; completed population-based and sample-survey research on the health of Gulf War veterans (United States, Canada, and the United Kingdom); VA and DoD health databases; the reports of the PAC, the GAO, and other IOM committees; and books and articles describing and evaluating approaches and instruments for measuring health status.

Questions Regarding the Health of Gulf War Veterans

The first task before the committee was to identify questions regarding the health of Gulf War veterans. A review of sources such as congressional testimony, GAO reports, and presentations to the committee pointed to several key groups and their questions. Some of these questions can be addressed by research, but others are in the realm of policy.

Gulf War veterans want to know how many of them are ill and why. They are concerned about their medically unexplained symptoms. They want to know whether they will get better or not, and they want reassurance that the government is trying to help them. Veterans also want to know whether DoD and VA are providing them with the treatments they need in an appropriate and timely manner.

VA and DoD are also asking questions. They want to know whether the health of ill veterans is getting better, worse, or remaining constant; how to best track the health of Gulf War veterans over time, especially for ill-defined or undiagnosed conditions; and what scientific studies should be conducted to resolve areas of continued scientific uncertainty related to health outcomes and treatment efficacy.

Questions from Congress and GAO concern how many veterans are ill and their clinical progress over time, that is, whether ill veterans are better or worse now than when first examined. They want to know whether veterans are receiving appropriate, effective, high-quality care; about plans for collecting longitudinal information on the health of ill veterans; whether the number of ill veterans is increasing; and whether these veterans are ill because of something that happened in the Gulf.

Furthermore, many individuals and groups are now beginning to ask if these questions apply only to Gulf War veterans or if they also apply to the veterans of any conflict.

The committee concluded that no one study can answer all the questions, but that it is possible to design a study that will measure the health of Gulf War veterans and permit comparisons with the public in general, veterans who were not deployed to a conflict, and veterans of other conflicts. This report discusses the deliberations and recommendations of the committee, including a research portfolio designed to help answer some of the questions being asked about the health of Gulf War veterans. The research portfolio and its centerpiece, a prospective cohort study, are intended to (1) monitor the current and future health status of Gulf War veterans; (2) identify the extent to which there are differences in health between Gulf War veterans and other groups who did not serve in the Gulf War; and (3) provide information that can be used to generate hypotheses about why any such differences exist.

Specifically, the study designed by the committee is intended to answer the following questions:

1. How healthy are Gulf War veterans?
2. In what ways does the health of Gulf War veterans change over time?
3. Now and in the future, how does the health of Gulf War veterans compare with that of

    - the general population;
    - persons in the military at the time of the Gulf War but not deployed;
    - persons in the military at the time of the Gulf War who were deployed to nonconflict areas; and
    - persons in the military deployed to other conflicts, such as Bosnia and Somalia?

4. What individual and environmental characteristics are associated with observed differences in health between Gulf War veterans and comparison groups?

DEFINING GULF WAR VETERANS

Various definitions related to the Gulf War are relevant to this committee's work. The definitions used by the VA and DoD differ, making it essential that the committee specify those that it has adopted.

The VA defines the Gulf War Era as beginning on August 2, 1990, and continuing to the present. Other definitions used by the VA include

- Gulf War Conflict Veteran—discharged from active duty status on or after August 2, 1990, and with service in the Southwest Asia theater of operations and with in-theater service occurring between August 2, 1990, and July 31, 1991.
- Gulf War Theater Veteran—discharged from active duty status on or after August 2, 1990, and with service in the Southwest Asia theater of operations.
- Gulf War Era Veteran—discharged from active duty status on or after August 2, 1990.

The definitions generally followed by DoD are those published in the final report of the Presidential Advisory Committee on Gulf War Veterans' Illnesses (1996a).

- Gulf War theater of operations is defined as including the Persian Gulf, Kuwait, Iraq, Saudi Arabia, Red Sea, Gulf of Oman, Gulf of Aden, northern portion of the Arabian Sea, Oman, Bahrain, Qatar, and the United Arab Emirates.
- Gulf War veteran is a person who served on active duty in the Gulf War theater of operations anytime during the period from August 2, 1990 (Iraq invaded Kuwait) to June 13, 1991 (last U.S. service member who participated in the ground war returned to the United States).
- Gulf War era veteran is defined as anyone who served on active duty anywhere other than the Gulf War theater of operations during the August 2, 1990, through June 13, 1991, time frame.

For purposes of this report, the committee has followed the PAC definition of a Gulf War veteran, that is, any person who served on active duty in the Gulf War theater between August 2, 1990, and June 13, 1991. Thus, the population of Gulf War "veterans" includes individuals who were on active duty at the time of Iraq's invasion of Kuwait as well as individuals called up for active duty from reserve and National Guard units. It may include individuals who remain on active duty or continue to serve in reserve units or the National Guard. Similarly, "veterans" of other conflicts (e.g., Bosnia, Somalia), who are of interest as a comparison group, are persons who served on active duty in a designated area during a specified period of time, regardless of their current service status.

STRUCTURE OF THE REPORT

Chapter 2 reviews some of the numerous research studies that have examined specific aspects of the health of Gulf War veterans. Chapter 3 discusses concepts of health and approaches used to measure it. Chapter 4 describes the committee-developed portfolio of research studies for coordinating research efforts aimed at addressing questions about the health of Gulf War veterans.

Chapter 5 details the components of the prospective cohort study designed by the committee, including its general design, sampling and scheduling issues, selection of survey modes and instruments, cost, ethical issues, and oversight responsibilities. The final chapter presents the committee's conclusions regarding the need to establish a means for measuring the health of Gulf War veterans and assessing changes over time.

2

Studies of the Health of Gulf War Veterans

INTRODUCTION

The second component of the charge to this committee is to identify issues to address in the development of study designs and methods used to answer identified questions regarding the health of Gulf War veterans. The committee began by reviewing and evaluating previous studies to determine what they can tell us about the health of these veterans. In this chapter we first review published studies that examine Gulf War veterans' health. This is followed by a discussion of the methodological strengths and limitations of the research to date. Finally, this chapter describes the evidence available to date on adverse health effects related to service in the Gulf War and the basis that current evidence provides for the design of the study described in Chapter 5.

A rich literature exists on adverse health effects of military conflicts, including effects attributed to exposure to mustard gas and Agent Orange. Additionally, a recent article by Hyams and colleagues (1996) summarized reports of poorly understood, multisymptom clusters recorded in conflicts dating back to at least the Civil War. They found reports of symptoms include fatigue, shortness of breath, headache, sleep disturbance, forgetfulness, and impaired concentration. No single etiologic entity has been discovered to account for these symptoms in conflicts prior to the Gulf War, and no generally accepted diagnostic label or set of clear clinical criteria has been developed out of earlier conflicts to use in the assessment of health problems among Gulf War veterans.

For earlier conflicts, various stress-related or psychogenic explanations were put forward to account for the cluster of symptoms observed among veterans. Shell shock, combat fatigue, irritable heart, or effort syndrome are all terms that have been used to both label and possibly explain the observed symptomatology. None of the labels was particularly useful in leading to clear diagnostic criteria, effective preventive interventions, or effective clinical treatments. There was a clear stigma of weakness associated with many of the labels and associated explanatory theories, so they were understandably not widely embraced by veterans or their families.

The Vietnam experience contributed case criteria and a label for posttraumatic stress disorder (PTSD). The disorder itself was not new, but the recognition of it and the eventual incorporation of PTSD into the standard medical and psychiatric diagnostic coding systems (ICD-9 and DSM-III [*Diagnostic and Statistical Manual*], respectively) date from the Vietnam era. Additionally, concerns regarding health effects related to Gulf War exposures are similar to the concerns that emerged regarding Agent Orange exposure for Vietnam veterans.

Health problems in Gulf War veterans, then, are studied with the recognition of their potential similarity to problems in other conflicts, but not necessarily with the acceptance of prior causal explanations. Most studies take Gulf War deployment (yes/no) as the measure of exposure, because the set of potential causal factors for postwar health problems was not well understood at the time of the conflict and is not much more clearly understood today. A few studies, however, focus on a specific chemical or biological exposure (again, usually yes/no rather than a graded scale). The studies discussed below typically are designed to detect an unusually high frequency or severity of health problems in a group of veterans who served in the Gulf, compared to either veterans of the same period who were not deployed or a more general population who did not serve.

SPECIFIC STUDIES OF GULF WAR VETERANS

Mortality

Two studies focused on mortality among Gulf War-deployed veterans compared to similar veterans not deployed to the Gulf. The study by Kang and Bullman (1996) used the Beneficiary Identification and Records Locator System (BIRLS) of the Department of Veterans Affairs (VA) to track deaths in nearly the entire population of deployed Gulf War veterans. Death rates in the deployed group were compared with the rate in a similar-sized control group of active duty, National Guard, and reserve personnel who served during the Gulf War period but were not deployed. Deaths for any cause were tracked through September 1993. Writer et al. (1996) compared deaths occurring during the conflict among the population of deployed veterans and among a large population of nondeployed veterans in the 1990–1991 period. Data on deaths were obtained from Report of Casualty forms (DD Form 1300).

The Kang and Bullman study is perhaps more directly relevant to the committee's charge, because it was designed to detect excess mortality among deployed veterans in at least the immediate post war period. No such excess was found, except for deaths attributed to accidents. A similar finding was also noted among Vietnam veterans and may reflect a set of risk-taking phenomena among veterans returning from conflict rather than a chemical, biological, or psychological exposure leading to physiological change. During the 1998 meeting of

Federal Investigators on Gulf War Illness, Kang presented similar results based on an updated analysis of mortality data through 1996. The Writer study also found no excess mortality among deployed veterans once deaths directly related to combat were excluded.

Data from these studies indicate no measurable increase in mortality, other than accidents, either during the Gulf War itself or in the period through 1996.

Disease Incidence and Prevalence

Some studies have specifically focused on incidence of diagnosed disease (using ICD-9 or ICD-10 labels and codes) in cohorts of Gulf War veterans. Most notably, diagnostic data are available from the VA's Persian Gulf Registry (DVA, 1999) and the DoD's Comprehensive Clinical Evaluation Program (IOM, 1998). Tables 2-1 and 2-2 present the distribution of diagnoses by major disease categories. It is important to note that data collection began in 1992 but in 1996 a revised code sheet was implemented. This revision makes it impossible to aggregate complaint, symptom, and diagnostic data for the total population of veterans whose data are recorded in the VA registry. The revision does, however, allow for a greatly expanded self-report exposure history questionnaire. In addition, although the original recording form allowed up to 3 symptoms and up to 3 diagnoses to be listed, the revised form allows expanded recording of up to 10 symptoms and up to 10 diagnoses. Although absolute prevalence rates are different in the two series, the relative distribution across major categories is very similar, with diseases of the musculoskeletal system and mental disorders being the two leading categories in both sets of patients. Because the data come from self-selected case series (i.e., veterans who presented for care in special Gulf War health assessment and treatment programs), the relative prevalence of various diagnoses is informative, but no conclusions can be drawn about risk for various diseases among Gulf War veterans versus nondeployed veterans or a civilian population with similar demographic characteristics.

Data obtained in a similar manner were reported in a study of British Gulf War veterans by Coker and colleagues (1999). They found that 59 percent of veterans who presented for care at a Ministry of Defense Gulf War Assessment Programme had more than one diagnosis. The five most common categories of diagnosis (in order of prevalence) were diseases of the musculoskeletal system, mental or behavioral disorders, chronic fatigue syndrome, diseases of the respiratory system (mainly asthma), and diseases of the digestive system. Twelve percent of the participants had a PTSD diagnosis. Because the study did not seek to compare the symptom experience of deployed and nondeployed veterans, the prevalence of diagnoses and the relative distribution of diagnoses across organ systems are useful as descriptions of the problems found in a selected case series of program participants, but these data are not generalizable to all British Gulf War veterans or Gulf War veterans of other nationalities.

Two studies focused specifically on diagnosis of birth defects among infants born to U.S. Gulf War veterans. Cowan and colleagues (1997) examined the incidence of diagnosed birth defects in children born in the 1991–1993 time period to Gulf War veterans remaining on active duty. Records for all live births at 135 U.S. military hospitals were examined. Birth defects were identified from medical records that included any ICD-9-CM codes in the range 740–759, plus codes for neoplasms or hereditary diseases. The study included 33,998 infants born to veterans deployed to the Gulf and 41,463 infants born to nondeployed veterans. The risk of birth defects was the same for deployed as for nondeployed veterans (odds ratio .97 for male veterans, 1.07 for female veterans), and rates for both groups were very similar to rates in the civilian population. Length of service in the Gulf was not associated with risk of birth defects, and a separate analysis of severe birth defects also showed no association with service in the Gulf.

Araneta and colleagues (1997) found among newborns of Gulf War veterans a relative risk of 3.0 for Goldenhar Syndrome (a rare abnormality of facial structure), but the increased risk was not statistically significant. A small number of cases produced very broad confidence intervals (0.63–20.6) around this relative risk estimate, indicating that it could have been produced by chance alone.

TABLE 2-1 Distribution of Diagnoses for Participants in the VA Persian Gulf Registry (PGR)[a]

	Original Code Sheet[b] (n = 53,935)		Revised Code Sheet[c] (n = 19,721)	
Diagnosis	No.	%	No.	%
Musculoskeletal and connective tissue	13,299	25.5	7,286	36.9
Mental disorders	7,995	15.1	6,887	34.9
Respiratory system	7,540	14.3	3,626	18.4
Skin and subcutaneous tissue	7,144	13.5	3,813	19.3
Digestive system	6,028	11.4	3,451	17.5
Nervous system	4,398	8.3	3,441	17.4
Circulatory system	3,747	7.1	2,083	10.6
Infectious diseases	3,715	7.0	1,785	9.1
Injury and poisoning	2,485	4.7	2,020	10.2
Genitourinary system	1,774	3.4	1,126	5.7
Neoplasm	232	0.4	149	0.8

[a]As of February 1999; data were prepared by the DVA Environmental Epidemiology Service.

[b]Data were collected using the original code sheet implemented in 1992 and used until 1996.

[c]Data were collected using the revised code sheet implemented in 1996.

TABLE 2-2 Distribution of Diagnoses for Participants in the DoD CCEP

	Original Code Sheet[a] (n = 18,495)		Revised Code Sheet[b] (n = 10,242)	
Diagnosis	No.	%	No.	%
Musculoskeletal	3,419	18.8	2,140	22.1
Psychoses and mental disorders	3,385	18.6	1,645	17.0
Signs, symptoms, and ill-defined conditions	3,236	17.8	1,721	17.7
Healthy	1,712	9.4	583	6.0
Respiratory system	1,226	6.7	597	6.2
Digestive system	1,141	6.3	626	6.5
Skin and subcutaneous tissue	1,152	6.3	526	5.8
Nervous system and sensory organs	1,047	5.8	512	5.3
Infectious diseases	457	2.5	268	2.8
Circulatory system	383	2.1	302	3.1
Endocrine/nutritional metabolic diseases	366	2.0	248	2.6
Genitourinary system	226	1.2	140	1.4
Injury and poisoning	140	0.8	117	1.2
Neoplasms	153	0.8	102	1.1
Blood and blood-forming organs	105	0.6	102	0.7

[a]Data were collected from inception until April 1996 using the original code sheet.
[b]Data were collected from April 1996 through March 1998 using the revised code sheet.

Signs, Symptoms, and Specific Impairments

Preliminary reports on unusual clusters of symptoms among Gulf War veterans began to appear in the media in late 1992, and publications in peer-reviewed journals began to appear in 1993. Perconte (1993) and Southwick (1993) reported on what they described as psychological and war stress symptoms and trauma-related symptoms among relatively small numbers of Army, Marine, and National Guard reservists who had been deployed to the Gulf. The analysis focused on mental health issues in general, and symptoms of PTSD in particular. Both studies noted relatively high prevalence of PTSD symptoms among deployed veterans, and the Perconte study also noted higher levels of depression and general psychiatric symptomatology in deployed veterans than in a comparison group of nondeployed veterans.

Sostek and colleagues (1996) found a higher prevalence of gastrointestinal (GI) complaints (e.g., abdominal pain, gas, loose stools, and nausea and vomiting) among members of a National Guard unit who had been deployed to the Gulf than among members of the same unit who had not been deployed. The deployed veterans had GI symptoms while serving in the Gulf that continued after return to the United States. These data were obtained in a self-report survey focusing on GI symptoms. No clinical examinations were performed, and no attempt was made to provide clinical diagnoses for the symptoms reported.

In a larger study of deployed veterans, Stretch and colleagues (1995) surveyed all active duty and reserve personnel in Army, Navy, Marine, and Air Force units in Hawaii and Pennsylvania. Questionnaires were distributed to approximately 16,000 individuals through their units; responses were received from 4,334, of whom 1,739 had been deployed to the Gulf. Deployed veterans were two to three times as likely as nondeployed veterans to report a variety of symptoms. The investigators concluded that the symptoms were related to deployment and that the physical symptoms were more strongly associated with deployment than psychological health measures (Stretch et al., 1996a). An algorithm based on a cluster of symptoms (e.g., fatigue, malaise, rash, headache, and respiratory symptoms) that might characterize a possible "Gulf War syndrome" identified 178 deployed veterans, as well as 55 nondeployed veterans. The investigators noted a four-fold difference in PTSD prevalence rates between deployed veterans (9.2%) and nondeployed veterans (2.1%) (Stretch et al., 1996b).

A study by Proctor and colleagues (1998) compared the symptom experience of Gulf War veterans from New England and New Orleans to that of members of a single National Guard unit who had been sent to Germany during the same time period. Symptoms from the 52-item Expanded Health Symptom Checklist were grouped into nine body-system clusters; scores for the body-system groupings were the sum of 0–4 frequency ratings for either the three symptoms in the group or, if more than three symptoms were included in a body-system group, the three most representative symptoms (determined by expert judgment). Veterans also reported on exposure to various toxic substances, biological agents, and combat stressors, and were assessed for PTSD using two different instruments.

Gulf War veterans reported a higher level of symptom experience than those deployed to Germany on all but one of the symptom groups. The largest differences between the Gulf-deployed and Germany-deployed cohorts (in terms of odds ratios) were for dermatological, neuropsychological, and gastrointestinal symptoms. Rates of PTSD were also higher in Gulf-deployed veterans, and their self-reported general health status scores were lower. The meaning of these differences is uncertain, though, as individuals from the New England and New Orleans cohorts were selected for study based on symptom reports in 1992–1993 while the Germany-deployed cohort was not.

Even though associations were found between specific wartime exposures and presence of symptoms, there is uncertainty about the significance of this association because the data on exposures were from self-reports several years after the fact.

Another approach to defining Gulf War illness was reported by Haley, Kurt, and Horn (1997b), based on a study of 249 U.S. naval reservists from a single battalion that had been called to active service in the Gulf. Members were from five southeastern states. Study participants completed surveys in special face-to-face sessions conducted in two cities in December 1994 and January 1995. The investigators used factor analysis to identify underlying relationships among the individual symptoms and identified six possible clusters. The investigators did not find an association between psychological factors and having one or more of the six clusters; they concluded that the six symptom clusters could represent six syndromes or six variants on a single syndrome. It is important to note that the participants in this study were members of a single unit of Gulf War veterans that had already been extensively studied. Additionally, the response rate was low and there was no control group.

A second report from this group (Haley and Kurt, 1997) on the same 249 veterans noted associations between specific types of chemical exposures and specific symptom clusters. However, the results are suggestive rather than conclusive given the limitations which include relatively small sample size, conceptual overlap among some of the lists of symptoms in the six symptom clusters, and reliance on retrospective self-report data on possible chemical exposures.

A third study by Haley and colleagues (1997a) reported on a series of neurological tests in a small group (23) of veterans who met their case definition for a Gulf War-related symptom cluster compared to 10 controls who were deployed to the Gulf but did not meet case criteria and 10 veteran controls who were not deployed. A series of laboratory investigations on neurologic function was performed, and several specific dysfunctions were noted with more frequency in the case group than in either of the control groups. The authors concluded that the three factors identified in their prior studies represented three variants of a more general neurologic injury related to service in the Gulf (Haley et al., 1997a). No specific diagnoses could be made, and no specific wartime exposures were identified as causal factors for the neurologic dysfunctions. Some demographic factors (e.g., age) were associated with the symptom cluster experience. Unfortunately, there is no clear definition for a "case," rather the authors studied a group of subjects scoring high on these symptom factors for a broad range of neurologic tests, compared to groups of asymptomatic controls. Other limitations to this study include concern about the use of pooled scores for comparisons and multiple comparisons.

Fukuda et al. (1998) conducted a factor analysis study of symptoms on a much larger sample of Air Force and Air National Guard veterans (3,723) from four units in Pennsylvania. A detailed set of survey, clinical evaluation, and laboratory test data was assembled. Approximately one-third of the sample had been deployed to the Gulf, so the investigators had the opportunity to compare symptom experience, and clinical, and laboratory findings for the deployed and nondeployed groups.

The factor analysis originally identified 10 possible factors, but confirmatory factor analysis produced two factors accounting for most of the variance in symptom experience: mood-cognition-fatigue and musculoskeletal. A total of 10 symptoms were included in the two factors. A third factor was identified based on clinical experience with chronic fatigue. An individual was labeled a case if he or she had one chronic symptom from two of these three categories. Although deployed veterans were significantly more likely than nondeployed veterans to meet case criteria, perhaps the most notable finding from this study was that approximately 15% of the nondeployed also met the case definition. Clinical and laboratory findings were generally not different between the deployed and nondeployed groups, and no specific characteristics of service in the Gulf were predictive of the multisymptom illness.

Perhaps the strongest study on Gulf War veterans' experience of symptoms related to deployment in the Gulf is the "Iowa Study," a population-based study of all military personnel who listed Iowa as the home of record at enlistment (Iowa Persian Gulf Study Group, 1997). A total of 29,000 individuals were potential study participants; a stratified random sample was created to study approximately 750 individuals in each of four groups: Gulf War regular military; Gulf War National Guard or reserve; non-Gulf War regular military; and non-Gulf War National Guard or reserve. Gulf War status was based on deployment to the Gulf War theater. Other stratification variables (e.g., age, sex) were built into the sampling design; 4,886 individuals were selected for possible inclusion in the study and 3,695 were actually contacted by telephone and interviewed. Topics in the interview included symptom experience, wartime exposures, occurrence of injuries or specific illnesses (e.g., cancer), and PTSD.

The investigators developed a priori definitions for a series of health outcomes of interest, such as cognitive dysfunction, fibromyalgia, depression, bronchitis, asthma, and anxiety. They used structured instruments such as the PRIME-MD and the SF-36 to arrive at these outcomes. Not only did those deployed to the Gulf report symptoms significantly more often, but most of the medical and psychological a priori outcomes were also elevated in the deployed group. The greatest differences between deployed and nondeployed respondents were seen for symptoms of cognitive dysfunction, fibromyalgia, and depression. The associations between exposures and the outcomes of symptoms of fibromyalgia, depression, and cognitive dysfunction were with "categories" of exposure thus, for example, associations were seen between one or more petrochemical exposures, and each of these three major outcomes.

Similar findings about higher prevalence of clusters of cognitive and other symptoms for deployed veterans were reported by Wolfe et al. (1998) in a study of approximately 3,000 veterans from New England, by Pierce (1997) in a study of 525 women veterans, and by Unwin and colleagues (1999) in a study of 8,000 veterans from the United Kingdom. The Unwin study, a mail-out survey assessing symptoms, is noteworthy for its ability to compare veterans who served in the Gulf War to veterans from the same era who served in Bosnia and those who served in the Gulf War era but were not deployed to either conflict. Service in the Gulf War was associated with significantly higher rates of symptom experience than service in Bosnia or nondeployment. The authors did report an association between vaccinations and the multisymptom cluster identified by Fukuda et al. (1998).

A separate study by Ismail and colleagues (1999) used factor analysis of symptom data from United Kingdom servicemen deployed to the Gulf. This report identified 10 possible factors, which were then reduced to three by confirmatory factor analysis. The three factors were labeled mood-cognition, respiratory system, and peripheral nervous system. The authors noted some similarities between their factor structure and that reported by Haley (1997a); however, they also noted significant differences, attributable perhaps to differences in sample definitions and wartime experiences, or to somewhat different original lists of symptoms being used for the analyses. The absence of a musculoskeletal factor in the Unwin group's analyses is perhaps the most noteworthy difference.

The survey of Canadian Gulf War veterans (Goss Gilroy, 1998) showed significantly higher rates of self-reported chronic conditions and symptoms of a variety of conditions (fibromyalgia, cognitive dysfunction, PTSD, depression, bronchitis, asthma, multiple chemical sensitivity, and anxiety) among deployed veterans compared to controls. Odds ratios for various conditions ranged from 1.35 (respiratory disease) to 5.27 (chronic fatigue). Rates of self-reported chronic conditions were also generally higher in both the deployed and control veterans than in an age-matched general population sample of the Ontario Health Survey.

The Canadian Study and the Iowa Study produced very similar results. Some of the differences in prevalence between the two studies may be attributable to the use of a priori outcomes in the Iowa Study and differences in deployment of the Canadian veterans who were more often in supporting, non-frontline areas.

Most of the results discussed in this section come from self-report data obtained through surveys or telephone interviews. Some questions can be raised about the accuracy or validity of reports, particularly reports about specific exposures during the Gulf War that occurred several years before the survey and were not well documented or perhaps even noted at the time. Nevertheless, the consistency of findings of higher prevalence of cognitive, musculoskeletal, and energy/fatigue symptoms among veterans deployed to the Gulf is striking. Given the subjective nature of most of the symptoms some psychogenic origin for the complaints may exist. However, the consistency of reports across services, parts of the country, and across countries suggests that these are very real experiences, even if not fully understood.

Gulf War veterans who participated in the CCEP program completed detailed symptom surveys. A report by Kroenke (1998) described the experience of more than 18,000 veterans who participated in the CCEP program through April 1996. The same general set of symptoms identified in the Iowa studies and most other studies appears here: fatigue, cognitive problems, muscle and joint pains, rash, and so on. Because the study reports on self-selected program participants, drawing any inferences about relative frequency of symptom experience as a result of deployment to the Gulf is not possible. The similarity of distribution of symptoms in this large group to that reported in more population-based studies is striking. No specific exposures during the war were associated with individual symptoms, but individuals who reported exposures to more things also reported more symptoms.

Three other studies related some aspect of symptom experience to either specific aspects of Gulf War experience or demographic factors. Morgan et al. (1998) found that PTSD symptoms among deployed veterans were more severe during months associated with the anniversary of particularly stressful events such as seeing a fellow soldier killed or being in a missile attack. Sutker (1994) noted a particularly high prevalence of PTSD and other psychological symptoms among veterans who had been assigned to graves registration duties during the Gulf War. Sutker (1995) also reported higher levels of psychological distress (assessed through a combination of psychological tests) among women and minority veterans who had served in the Gulf.

On the basis of all these studies it appears that veterans who served in the Gulf are more likely than their nondeployed comrades or civilians to experience a set of symptoms that include cognitive, musculoskeletal, and energy/fatigue elements. However, no study has yet included a representative sample of the entire population of Gulf War veterans with appropriate comparison groups. In some cases, the symptoms are severe enough to be totally debilitating. Not all veterans experience the same cluster of symptoms; therefore, assuming a single underlying pathology or single cause for the complaints would not be appropriate. Despite intensive study in a number of large cohorts, it is impossible to say which exposure(s) in the Gulf War are associated with the symptoms being reported and what the underlying mechanism(s) may be.

Functional Status and Well-Being

The study of Canadian Gulf War veterans (Goss Gilroy, 1998) indicated a slightly higher proportion of bed days due to health in the 2 weeks prior to survey administration in deployed veterans compared to controls. Deployed veterans also reported more days in the past 2 weeks with "activity cut down" due to health (27% versus 15% in controls). Deployed veterans were more likely to respond in the "less favorable" half of a scale of general health than were controls.

The Iowa Study (1997) included the RAND Short Form-36 (SF-36) questionnaire in its data collection battery. Deployed veterans reported statistically significant lower scores than did the nondeployed on all eight subscales of the SF-36. Absolute differences on a 0–100 scale were smaller for physical functioning (2 points) and social functioning (3 points), but were larger for general health (7 points) and vitality (8 points). Because differences of even 1 point or a fraction of a point have been considered meaningful in other large population studies (Allen, 1997), a 7- or 8-point difference is substantively, as well as statistically, significant.

The Fukuda study of Air Force and Air National Guard veterans used SF-36 questionnaires for the subset of participants who volunteered for a clinical evaluation (Fukuda et al., 1998). The clinical evaluation was offered only to members of the index unit (the National Guard unit that was deployed to the Gulf), so one cannot draw any conclusion about any effect of deployment on SF-36 scores. It was noted, though, that SF-36 scores were significantly lower among those veterans who met criteria as a severe case of multisymptom illness (according to symptom criteria) than among those labeled as mild-to-moderate cases, who in turn had lower scores than noncases.

Health Care Utilization

The largest and most significant study of health care utilization, specifically hospitalization, among Gulf War veterans was conducted by Gray and colleagues (1996) at the Naval Health Research Center in San Diego. They studied the predeployment and postdeployment hospitalization experience of essentially all deployed military personnel who remained on active duty through 1993 to that of nondeployed military personnel for the same time period. Data for the study came from the Defense Manpower Data Center (used to identify eligible individuals) and discharge summary data for all DoD hospitals. The study sample was very large; approximately 1 million individuals were included in follow-up analyses for 1991, 1992, and 1993.

The study tested the hypothesis that deployed veterans would be admitted to DoD hospitals more frequently than their nondeployed counterparts, either for any cause or for specific conditions suspected of being caused by exposures during the Gulf War. Examples of the specific conditions included infectious and parasitic diseases, cancers, nervous system diseases, and musculoskeletal diseases.

The basic finding of the study was that risk of hospitalization for deployed veterans in the immediate postwar period was no different from that of nondeployed veterans, either for any cause or for virtually all the specific disease conditions. The one or two significant findings were for 1 year only and did not appear in previous or subsequent years (e.g., more admissions for disease of the blood in 1992 but not in 1991 or 1993); moreover, these findings were not always in the predicted direction (deployed veterans were significantly *less* likely to be admitted for ill-defined conditions in 1991 and 1992).

A healthy soldier effect was noted: admission rates for deployed veterans were significantly *lower* than those for nondeployed veterans in the 2 years prior to the Gulf War, and the largest difference in the period occurred immediately before the war. The effect was not found before 1990, however, suggesting that those deployed to the Gulf were perhaps in better health at that time than their nondeployed counterparts, but were not necessarily healthier due to stable causes that would predict hospitalization rates for years into the future.

The study was limited because of its focus on active duty personnel; conceivably, those suffering from Gulf War-related symptoms might leave active duty voluntarily or take a medical discharge. Hospitalizations for that group would appear in VA or private sector databases but not in the DoD database.

Despite this limitation, the study would have been able to detect any marked increase in hospitalizations among those who did stay on active duty (and perhaps would have been admitted at least once before going into any sort of medical discharge status); no such effect was seen.

A study from the same data set on hospitalizations that focused on unexplained symptoms or conditions showed a slight excess of such admissions for deployed veterans compared to nondeployed veterans (Knoke and Grey, 1998). The exclusion of veterans who had been admitted through the CCEP for evaluation not only eliminated the difference, this analysis suggested a lower rate of hospitalization for unexplained symptoms in the deployed group.

Summary of Findings from Studies to Date

The committee found that the large and growing literature on the health of Gulf War veterans supports both the conclusions that follow and the list of questions to be addressed by the prospective cohort study described in Chapter 5.

- Military personnel who served in the Gulf War have had a significantly higher risk (at least through 1996) of suffering one or more of a set of symptoms that include fatigue, memory loss, difficulty concentrating, pains in muscles and joints, and rashes. Other symptoms are noted with reduced frequency, but still may be experienced more often by deployed than nondeployed veterans.
- The symptoms range in severity from barely detectable to completely debilitating.
- No single accepted diagnosis or group of diagnoses has been identified that describes and explains this cluster of symptoms.
- There is no single exposure, or set of exposures, that has been shown conclusively to cause individual symptoms or clusters of symptoms. Although some statistical associations have been seen in some studies, they have not been confirmed in other studies or confirmed through laboratory tests that would establish a cause–effect connection in individual patients.
- No diseases included in the ICD-9-CM or ICD-10 classification systems have been shown to be more frequent in deployed or in nondeployed veterans with the exception of PTSD symptoms.
- Mortality among deployed veterans is not higher in general than mortality among nondeployed veterans, at least through 1993. Deaths due to accidents are higher among deployed veterans.
- Health-related quality of life, as measured through instruments such as the SF-36, is lower on average among deployed veterans than among nondeployed veterans.
- The natural course of symptom experience over time is not known, as no longitudinal studies of symptom experience have been conducted and reported in the literature.

LIMITATIONS OF PREVIOUS STUDIES

Although the studies to date have provided valuable information regarding the health problems experienced by some Gulf War veterans, these studies have important limitations in terms of assessing the health status of those veterans. One concern is that available data and studies are not representative of the entire population of U.S. Gulf War veterans. For example, the Gulf War veteran registries developed by DoD and VA include only those veterans who have chosen to participate. The health status of participating veterans may differ from that of other Gulf War veterans. The studies of hospitalizations (Gray et al., 1996; Knoke and Gray, 1998) and adverse birth outcomes (Araneta et al., 1997; Cowan et al., 1997) have been limited to personnel remaining on active duty and to events occurring in military hospitals. Health status or characteristics of active duty personnel could differ from those of personnel who have left active duty or who have been treated in nonmilitary hospitals. Moreover, economic and other non-health-related factors are likely to have a greater effect on the use of nonmilitary hospitals and health care services.

As noted in the review above, other studies have focused on individual service units or other subpopulations of Gulf War veterans (e.g., Fukuda et al., 1998; Haley et al., 1997 a,b,c; Pierce, 1997; Stretch et al., 1995). The deployment experience of specific units (e.g., location of deployment, tasks during deployment) or other features of a unit (e.g., service branch, reserve or National Guard versus active duty) may be distinctive and therefore not generalizable to a larger Gulf War veteran population. Even the Iowa Study (Iowa Persian Gulf Study Group, 1997), which drew a representative sample from the population of all veterans with Iowa as home of record at time of enlistment without regard to service unit, may not be generalizable to all Gulf War veterans because of factors specific to Iowa such as a small minority population.

Another factor that may affect the representativeness of the studies conducted so far is bias introduced by low participation rates or by participation rates that differ among subgroups (e.g., Fukuda et al., 1998; Haley et al., 1997a,b). Several factors may influence participation rates. For example, veterans still on active duty or attached to reserve or National Guard units may be easier to locate than those who have been discharged. Gulf War veterans on active duty may be reluctant to identify conditions that could lead to a change in their duty status, whereas those veterans who perceive that they have service-related health problems may have a greater incentive to participate in studies than those who do not have health problems.

The studies to date are also limited in their assessments of the health status of Gulf War veterans. Some studies have used very specific indicators such as mortality (Kang and Bullman, 1996) or hospitalizations (Gray et al.,1996; Knoke and Gray, 1998). Many conditions, however, may impair health, functional status, and other aspects of well-being without causing death or requiring hospitalization. Furthermore, because the published mortality analysis covers only about 30 months following the return of many deployed troops, those data reflect only the most severe health effects that Gulf War veterans might experience. A longer period of observation will be needed to detect changes in mortality related to health problems with a long latency (e.g., cancer) or with a more chronic course (e.g., multiple sclerosis). Other studies have focused on indicators such as the prevalence of reported symptoms. Although symptoms can signal health problems, determining their presence does not by itself provide enough information to assess whether those symptoms affect function or other aspects of health.

REMAINDER OF THIS REPORT

The work of this committee is directed toward designing a research portfolio that will address the knowledge gaps and methodological problems identified thus far. Our overall aims are to develop a conceptual and operational framework that will produce a population-based assessment of the nature and extent of health problems among Gulf War veterans, assess the impact of these health problems on veterans' health status, and monitor changes in the health status of these veterans over time. Such a framework will generate information necessary for policymaking, clinical decision making, and shared decision making between health care providers and patients. The remainder of the report addresses the committee's proposed study framework for making such assessments.

3

Measuring Health

As the discussion in Chapter 2 has shown, studies of the health of Gulf War veterans have provided convincing evidence of health concerns for some veterans. These studies have not, however, succeeded in identifying specific factors that may have caused those health problems or in establishing firm diagnoses in many cases. The work of this committee represents an effort to move ahead with a more general and prospectively oriented approach for ensuring systematic study of the current and future health of Gulf War veterans. This chapter reviews the key aspects of defining and measuring health that have served to guide the committee in developing recommendations for future research.

THE EVOLVING DEFINITION OF HEALTH

Early health status assessments relied only on mortality. In comparing populations, those with the lower mortality rates were considered healthier and "better off" than those with higher mortality rates. Infant mortality and mortality rates of cohorts with defined diseases or at older ages continue to be used as general indicators of population health status.

As basic survival became less problematic and chronic disease became more prevalent, assessments of health status began to include measures of morbidity reflected in rates of illness and injury. With increasing scientific and medical knowledge, results of biochemical tests, observed symptom rates, and statistics on use of health care services were employed as indirect measures of morbidity. In the mid-1950s, the United States government initiated major surveys designed to collect population-based data about illnesses and injuries, and their effects on levels of activity in the population.

Also in the 1950s, attention began to focus on the World Health Organization (WHO) definition of health as "a state of complete physical, mental, and social well being, and not merely the absence of disease or infirmity" (WHO, 1948). Health was no longer defined only in the negative, that is, as the absence of disease, and was recognized as much more than simply the state of *not* suffering from any designated undesirable condition (Evans and Stoddard, 1994).

For health status assessment, the result was an expansion in how health was measured. The new definition of health necessitated development of new indicators that would measure complete physical, mental, emotional, and social well being. A frequently used approach relies on measures of physical and psychological functioning, that is, an individual's functional status or the extent to which an individual can function normally and carry on his or her typical daily activities. Talcott Parsons (1958) described functional status an individual's effective performance of or ability to perform those roles, tasks, or activities that are valued, for example, going to work, playing sports, maintaining the house. Bowling's (1997:4) definition of functional status is similar: "the degree to which an individual is able to perform socially allocated roles free of physically (or mentally in the case of mental illness) related limitations." Thus, conditions that limit an individual's ability to perform usual roles or tasks are recognized as threats to health. However, Patrick and Erickson (1993) and Bowling (1997) both pointed out that functional status is only one of the components of health.

Social well being or social health is another component of the WHO definition of health. It is considered distinct from physical and mental health and may be viewed in terms of adjustment, social support, or the ability to perform normal roles in society (McDowell and Newell, 1996). Bowling (1997) reported that social health has been described in terms of the degree to which people function adequately as members of the community; socially healthy persons would be more able to cope with day-to-day challenges. Individual personalities can be influenced by the quality and quantity of interpersonal relationships, and lack of social integration may produce stress and decrease an individual's resources for dealing with it.

A broader concept, health-related quality of life, has been described by Patrick and Erickson (1993:22) as "the value assigned to duration of life as modified by the impairments, functional states, perceptions, and social opportunities that are influenced by disease, injury, treatment, or policy." McDowell and Newell (1996) included measures of physical, emotional, and social dimensions of health in their description of this concept, although they asserted that researchers have, for the most part, not clearly distinguished between quality of life measures and general health measures.

Ware and colleagues (1993) described health-related quality of life outcomes as those most directly affected by disease and treatment. These outcomes include behavioral functioning, perceived well being, social and role disability, and personal evaluations (perceptions) of health in general. Gold et al. (1996), following the approach of Patrick and Erickson and narrowing it to measures that integrate survival and health, used the concept of health-related quality of life to represent the values assigned to different health states. Despite slight differences in terminology, there is general consensus on the concepts and domains included in operational definitions of health-related quality of life.

As this brief review indicates, the definition of health and the indicators used to measure health have evolved over time. McDowell and Newell (1996:11) summarized this evolution as "a shift away from viewing health in terms of survival, through a phase of defining it in terms of freedom from disease, onward to an emphasis on the individual's ability to perform daily activities, and more recently to an emphasis on positive themes of happiness, social and emotional well being, and quality of life." With a broader definition of health, it is no longer adequate to rely on measures of mortality, illness, or surrogate measures of health such as hospitalizations. Also needed are health measures that can represent social, psychological, and physical well being to obtain a meaningful picture of the overall status of both individuals and populations.

CORE CONCEPTS OF HEALTH

Table 3-1 displays the five core concepts of health-related quality of life: death and duration of life; impairment; functional status (physical, psychological, and social); health perceptions; and opportunity (Patrick and Erickson, 1993). These concepts encompass both the quantity and quality of life and can be further differentiated into several domains. Domains are states, attitudes, behaviors, perceptions, and other spheres of action and thought; for instance, under the concept of impairment, symptoms and subjective complaints is the first domain. In assessing health, researchers must choose relevant domains and subdomains for measurement. The five main concepts of health, with selected domains and subdomains, are briefly described below.

Death and Duration of Life

The mortality-based measures most frequently used in prior studies of Gulf War veterans include the total death rate, condition-specific death rates, and infant mortality. Future studies could also use such measures as potential years of life lost, and remaining years of life at various ages, or could combine mortality with health status to form a composite measure such as years of healthy life, quality-adjusted life year, or disability-free life years. (Patrick and Erickson, 1993; Torrance and Feeny, 1989).

TABLE 3-1 Core Concepts and Domains of Health

Concepts and Domains	Definitions/Indicators
Death and Duration of Life	Mortality; survival; years of life lost
Impairment	
Symptoms and subjective complaints	Reports of physical and psychological symptoms, sensations, pain, health problems, or feelings not directly observable
Signs	Physical examination: observable evidence of defect or abnormality
Self-reported disease	Patient listing of medical conditions or impairments
Physiological measures	Laboratory data, records, and their clinical interpretation
Tissue alterations	Pathological evidence
Diagnoses	Clinical judgments after "all the evidence"
Functional Status	
Social function	
Limitations in usual roles	Acute or chronic limitations in usual social roles (major activities) of child, student, worker
Integration	Participation in the community
Contact	Interaction with others
Intimacy and sexual function	Perceived feelings of closeness; sexual activity and/or problems
Psychological function	
Affective	Psychological attitudes and behaviors, including distress and well being
Cognitive	Alertness; disorientation; problems in reasoning
Physical function	
Activity restrictions	Acute or chronic reduction in physical activity, mobility, self-care, sleep, communication
Fitness	Performance of activity with vigor and without excessive fatigue
Health Perceptions	
General health perceptions	Self-rating of health; health concern/worry
Satisfaction with health	Satisfaction with physical, psychological, social function
Opportunity	
Social or cultural disadvantage	Disadvantage because of health; stigma; societal reaction
Resilience	Capacity for health; ability to withstand stress; physiological reserves

SOURCE: Adapted from Patrick and Erickson, 1993.

Impairments[*]

The concept of impairment includes morbidity, which can be represented by measures such as the number of sick persons or cases of disease in relationship to a specific population. In the medical model of disease, morbidity includes both pathological processes that have not yet been recognized and those that have become evident. Diagnoses have typically been the focus of morbidity studies, but they are incomplete as a measure of health. Morbidity can be viewed more broadly as impairment, which encompasses both disease and any loss or abnormality of psychological, physiological, or anatomic structure or function.

Although health professionals apply varying etiological, anatomic, and physiological criteria for the evaluation of disease, at least six impairment domains can be distinguished: symptoms and subjective complaints, signs, self-reports of disease, physiological measures, tissue alterations, and diagnoses.

Symptoms and Subjective Complaints. Reports of physical and psychological symptoms, sensations, pain, health problems, or other feelings of abnormality are best known to the person who has them, and often are not directly observable by an interviewer or evaluator. Many condition and symptom checklists have been developed and are most useful for identifying condition-specific subgroups for analysis. Some of the most common health problems reported by Gulf War veterans (e.g., fatigue, difficulty concentrating, joint pain) fall in this domain.

Signs. In contrast to symptoms, which are usually best known to the person experiencing them, signs are objective, observable evidence of impairment. They are generally identified through physical examination.

Self-Reported Disease. In assessing the health of a population, health status surveys frequently ask respondents to report whether or not they have or have had heart disease, cancer, respiratory disease, or other specific diagnosable condition. Some of these reports will reflect information conveyed to the respondent by a physician. Other reports will be based on the respondents' own assignment of a diagnosis.

Physiological and Performance Measures. Clinicians use a vast number of physiological measures (e.g., blood count, glucose tolerance, forced expiratory flow, treadmill tests, cognitive tests) to detect abnormalities. Records from these tests may be single readings or an extensive series of measures, such as prolonged electrocardiograph monitoring to determine the efficacy or toxicity of a drug prescribed to treat cardiac rhythm disturbance.

Tissue Alterations. Alterations in body tissue may be detected by examination of tissue obtained either at autopsy or collected during a surgical procedure. For example, microscopic examination of heart tissue from a patient with

---

[*]Much has been written about how to define the concepts of impairment, disability, and functional status. Although space precludes reviewing the issues here, the 1991 IOM report *Disability in America* provides a discussion of these concepts.

atherosclerotic coronary heart disease can detect pathological processes such as necroses of cardiac muscle.

Diagnoses. Diagnoses represent categories of disease or medical conditions. They are arrived at through consideration of some mix of clinical history, observation, physical examination, reported symptoms, and physiological and performance measures.

Functional Status

Functional status has three major domains: physical, psychological, and social.

Physical Function. Physical function can be classified into two subdomains: activity restrictions and fitness. Commonly reported *activity restrictions* are restrictions in body movement (e.g., difficulty in walking or bending over), limitations in mobility (e.g., having to stay in bed or not being able to drive a car or use public transportation), or interference with self-care activities (e.g., not being able to bathe, dress, or eat without assistance). Commonly used indicators of restricted activity are work-loss days, school-loss days, days of restricted activity, and bed days. *Fitness* is assessed in terms of energy, endurance, speed, and the more positive nature of physical activity. Individual measures of fitness include such items as the ability to run the length of a football field or the speed with which one can walk 10 yards.

Psychological Function. Psychological function includes affective indicators of happiness, distress, morale, or mood and cognitive dimensions such as alertness, confusion, or impaired thought and concentration. The subdomain of *affective functioning* is assessed using general measures of distress, measures of specific mood states, and diagnosable conditions such as depression and anxiety. The subdomain of *cognitive functioning* refers to matters such as impaired thought and concentration, memory, and the ability to carry out intellectual functions essential to normal routines of living (e.g., remembering names and telephone numbers, performing tasks on the job).

Social Function. Social function refers to the individual's ability to maintain relationships with others in the context of work, neighborhood, and family. The broad concept is divided into several subdomains:

Limitations in Usual Roles or Major Activity. This domain includes the capacity for or performance of usual social roles (e.g., holding a job, going to school, parenting, managing a house, engaging in leisure pursuits, and maintaining relationships with friends).

Social Integration. This domain includes participation in the community through membership in social, civic, political, or religious organizations.

Social Contact. Social contact is assessed through indicators such as the frequency of visits with friends or relatives, the number of meetings and community activities attended, or types of social interaction.

Intimacy. This domain covers feelings of closeness and trust. Intimacy can be an important determinant of emotional well being in patients facing serious illness or death. Social contact and intimacy, of course, are not mutually exclusive; both can be provided by the same person or persons in a social network. Sexual function and dysfunction are included in this domain.

Health Perceptions

Individuals' subjective judgments of their health capture both the personal evaluative nature of health and the more positive aspects of the quality of life.

General Health Perceptions. Subjective health status is often assessed on the basis of a self-rating of health as excellent, very good, good, fair, or poor. Such judgments are considered ratings because they reflect individual differences in evaluating health. Global self-assessments, like more comprehensive and lengthy measures of general health perceptions, include the individual's evaluation of physiological, physical, psychological, and social well being and the effect of health on other aspects of life such as opportunity and respect.

Satisfaction with Health. The domain of satisfaction with health reflects the extent to which an individual's needs or aspirations are fulfilled.

Opportunity

Opportunity is defined as the potential for an optimal state of health or "being all that one can be." Capacity or potential can be represented through both the negative term, disadvantage and the positive term, resilience.

Social or Cultural Disadvantage. This domain of opportunity includes physical and social access to the environment, to education and training, and to employment. For example, patients with chronic renal failure who require dialysis may find it difficult to obtain or keep employment. Health disadvantage is assessed in relation to people who do not have a particular condition or significant illness.

Social and cultural disadvantage also includes the concepts of social reaction and stigma associated with a health-related condition. People with disabilities have described repeatedly the guarded references to the disability, changes in social relationships, and negative reactions of others to the differentness of disability.

Individual Resilience. Resilience is consistent with the concept of physiological reserve, which is the unused capacity that can be called upon in times of stress, crisis, or increased activity. Resilience, or the capacity for health, is most often measured by the ability to cope with or withstand stress or to maintain emotional equilibrium. This approach recognizes that people adjust differently to life situations altered by disease or treatment.

Application to Studies of Gulf War Veterans

Studies of Gulf War veterans conducted to date have measured specific components within the wide range of domains and subdomains of health-related quality of life just described. Most of these studies have focused on mortality and clinical assessment rather than on the veterans' own evaluation of their health. The studies of Iowa veterans (Iowa Persian Gulf Study Group, 1997) and Canadian veterans (Goss Gilroy, 1998), however, demonstrate the importance of the use of self-report data for understanding veterans' health, especially when the findings of these investigations can lead to policies for responding to Gulf War veterans' health concerns. Especially at this point after the Gulf War, strategies that depend largely on "population rates" of death or overt disease (especially established diagnoses), will not suffice. Programs that address a broad range of experience, perceptions, and health problems are needed for the remainder of the life expectancy of the Gulf War veterans.

The concepts just reviewed capture the diverse aspects of health outcomes in terms of health-related quality of life, which this committee believes must be taken into account in studying the current and future health of Gulf War veterans. Establishing a more complete understanding of the health of these veterans also requires consideration of a variety of other influences that represent correlates of health.

CORRELATES OF HEALTH

Health-related quality of life, measured at any specific point in time, is a function of a variety of influences that inevitably go beyond the specific exposures or risk factors (e.g., deployment to the Gulf) of greatest interest in a specific study. Characteristics of the individual and the environment, both past and present, influence health; the committee believes that a study of the health of Gulf War veterans should pay explicit attention to a variety of factors in addition to deployment to the Gulf or specific experiences there, that influence health and health-related quality of life.

In developing study designs and health assessment strategies, the committee has adopted a model of the correlates of health to guide identification of those characteristics of the individual and the environment that might be assessed in relation to health and general quality of life. This conceptual model is shown in Figure 3-1.

Individual Characteristics

Some health correlates are specific to an individual. These are represented on the bottom plane of Figure 3-1. *Biology* includes, most importantly, genetically based vulnerabilities, predisposition, and potentiating factors. *Life course* refers to the exposures actually encountered by the individual, including such factors as harmful chemical agents and life-event stressors. *Life-style and health behavior* includes individual activities and exposures, such as smoking, drinking, drug use, diet, and exercise—factors often referred to as risk behaviors. *Illness behavior* includes an individual's coping behaviors and information seeking in the face of a potential or actual health threat. *Personality and motivation* also refers to the behaviors of an individual, including his or her sense of locus of behavior control and proclivity for or aversion to risk taking. *Values and preferences* refers to the individual's cultural values and attitudes relating to health and illness and his or her concepts of the nature of health and illness.

Environmental Characteristics

Interacting with these individual-level variables will be the physical and social environments that influence individuals and groups of people, as depicted in the top plane of Figure 3.1. *Social and cultural* influences include role expectations, social networks, and dominant concepts of what is considered to be health and illness. *Economic and political* factors define the level and type of resources and constraints that permit or deter individuals from being able to respond to needs and desires. *Physical and geographic* factors have to do with the likelihood and frequency of protective or dangerous exposures such as climate, air quality, presence of environmental toxins, and crime. *Health and social care* include the public health infrastructure and personal health services available to respond to health threats and problems. Culture and the social environment help to define what is considered a health care *need* and how these needs are met.

Characteristics of Health-Related Quality of Life

Health-related quality of life, shown as the middle plane in Figure 3-1, is influenced by both individual and environmental factors that a person (or a population) experiences. The components of health-related quality of life were discussed in some detail earlier in this chapter. General quality of life, shown on the far right side of the figure, includes satisfaction with overall life as well as satisfaction with the environment and individual aspects of life. The arrows indicate the direction of interaction between the characteristics.

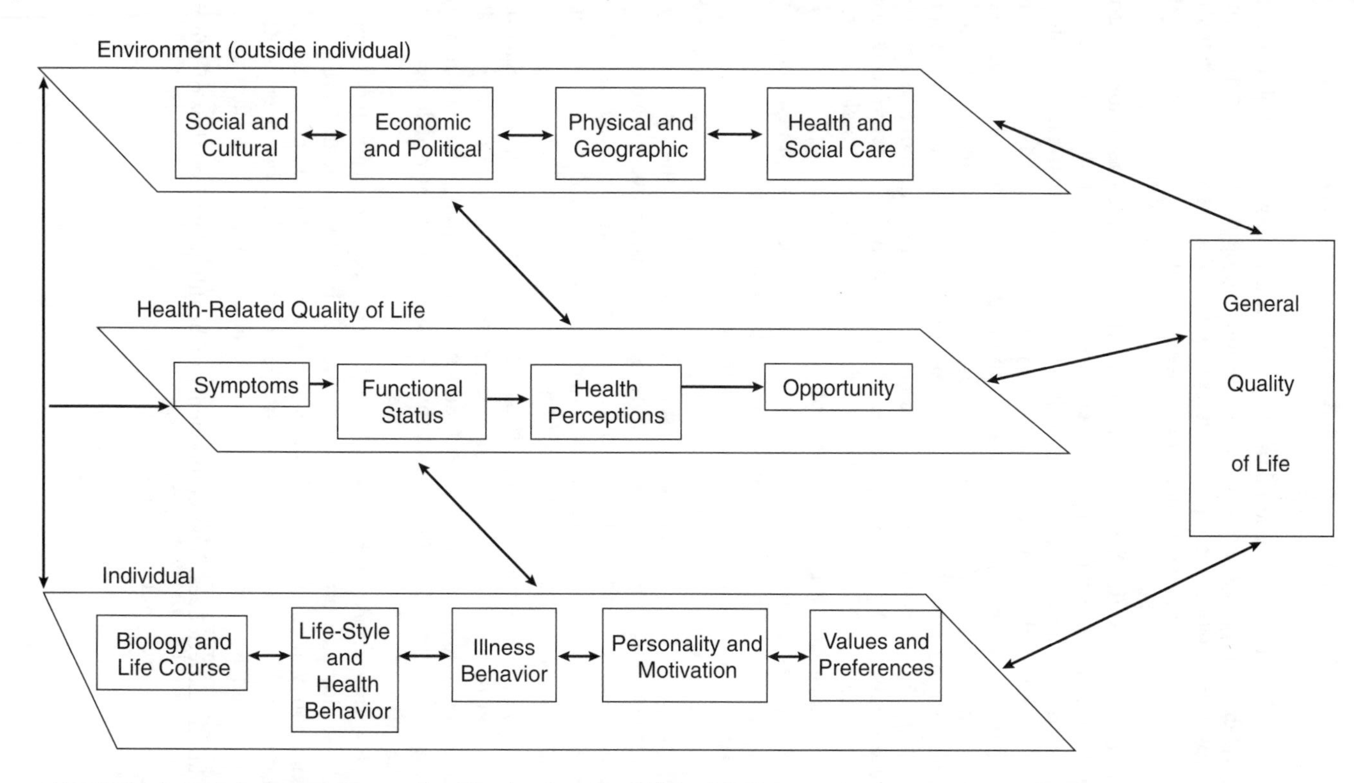

FIGURE 3-1 Model of determinants of health-related quality of life of Gulf War veterans.

MEASURING HEALTH-RELATED QUALITY OF LIFE

Health cannot be measured directly, as can the weight of an object. Instead, a number of variables are used as indicators of the overall concept of health. Different applications or types of studies require the use of health status measures with different measurement properties (Guyatt et al., 1993b; McDowell and Jenkinson, 1996; Scientific Advisory Committee, 1999). An assessment instrument that works well for one purpose or application or in one setting or population may not do so when applied for another purpose or in another setting or population. For example, for monitoring the health of a population, a single indicator such as the prevalence of limitation of major activity or self-rating of health may be used to gain information on trends over time. For epidemiological investigations, clinical trials, and studies of practice that are intended to select treatments and monitor individual patient response, more detailed or multidimensional health status information is needed. In addition to overall, or generic, data on individual functioning, investigators and clinicians may need to know about symptoms and specific areas of functioning, as well as side effects that may represent adverse consequences of treatments.

Regardless of the nature of a measure or its intended application, all measures used to assess health-related quality of life must be reliable and valid, and if evaluating change, they must be responsive. Reliable measures produce the same results when measurement is repeated. Valid measures are those that measure what they purport to measure, in this instance that means they must accurately reflect health-related quality of life. Responsive measures are those that are able to identify changes over time or in response to specific interventions. Similarly, the instruments used to obtain data on which such measures are based must meet accepted standards for reliability and validity.

When assessing health, one chooses to measure a sample of behaviors and traits; that is, one chooses domains relevant to the question of interest. Whether one chooses to combine different domains to calculate an overall summary measure will depend on the type of study or application for which the measure is needed.

For policy making, for example, one often needs a measure that summarizes multiple domains into an overall measure in order to compare across different health policy alternatives. Such a measure involves tradeoffs between the different domains in the decision process. It should be possible, however, to disaggregate a summary measure in such a way that analysts and decision-makers can assess the relative contribution of individual domains to the overall value assigned to health. In developing programs for Gulf War veterans, these points may be especially relevant for both policymakers and clinicians.

Features of Measures of Health-Related Quality of Life

The taxonomy of health measures shown in Table 3-2 addresses specific features of such measures related to their use for assessing health status and quality of life (Guyatt et al., 1993; Patrick and Erickson, 1993). The table includes summary comments on the strengths and weaknesses of each type of measure. Such information can be used to guide the selection of measures for specific applications. Although it is beyond the scope of this report to define in detail the characteristics of all possible measures of health that might be used, some definitions and a brief review of the characteristics of such measures will be helpful in interpreting later sections of this report.

The measures are classified according to (1) sources of data; (2) type of scores produced, which reflects the level of aggregation across concepts and domains; (3) range of populations and concepts/domains covered, including the different diseases, health conditions, and populations to be assessed and the breadth of concepts and domains included in the measure; and (4) the weighting system used in scoring items and in aggregating different domains.

Health measures can be differentiated on the basis of the means used to obtain data. *Physiological* measures (e.g., blood pressure, serum glucose) rely on direct testing or observation of biologic material such as blood or tissue samples or of physiological processes such as cardiac function. Physical *examination* provides clinically relevant information on an individual as observed by a physician or other trained personnel. *Performance-based* data are derived from results of defined tasks such as walking a specified distance. *Self-reports* are respondent-supplied information gathered using methods that include written questionnaires, interviews, internet administration, or computer-adapted testing.

Health measurement may be reported using any of several types of scores:

- *indicators:* a single number obtained from a single item;
- *indexes:* a single number summarizing multiple concepts or a classification of health-related quality of life;
- *profiles:* multiple numbers on the same metric; or
- *batteries:* multiple numbers on different metrics.

Most health measures discussed in this report are either profiles of scores or indexes in which a single number is derived from multiple items.

In terms of the range of populations or concepts covered, measures are characterized as generic or specific. *Generic* instruments cover a wide range of domains and are used primarily to compare across populations without regard to the specific condition of the person being assessed. *Specific* instruments focus on those aspects of health and quality of life that are of primary interest in terms of a population (e.g., older adults, children) or an individual; a disease (e.g., asthma or heart failure); a certain domain or function (e.g., sleep or sexual function); or a problem (e.g., pain). Specific measures can be seen as having the advantage of being relevant to clinicians and persons with specific conditions, but generic measures often have the advantage of permitting comparisons across varied groups, as is expected for Gulf War veterans, or against population norms calculated for numerous and/or large groups of respondents.

TABLE 3-2 Characteristics of Health Measures

Measure	Strengths	Weaknesses
Sources of Data		
Physiological measures, e.g., blood pressure	Physical trace can be stored or recorded	May be expensive and measures not available for many PGW conditions
Physical **examination**	Clinician-focused and diagnostically relevant	May not be reproducible and may be expensive
Performance-based, e.g., person asked to perform task and performance recorded	Can be standardized	Not available for many conditions or relevant to subjective reports
Self-reports	Directly from individuals affected	Difficult to establish validity as no gold standard or criterion exists
Type of Scores Produced		
Single **indicator** number	Represents global evaluation useful for population monitoring	May be difficult to interpret; trends may not be responsive to change
Single **index** number	Represents net impact Useful for cost-effectiveness	May not be possible to disaggregate contribution of domains to the overall score May not be responsive to change
Profile of interrelated scores	Single instrument Contribution of domains to overall score possible	May not be responsive to change Length may be a problem May not have overall score
Battery of independent scores	Wide range of relevant outcomes possible	Cannot relate different outcomes to common measurement scale May need to adjust for multiple comparisons May need to identify major outcome

Preference-based measures are those that are used primarily in economic and decision theory and that incorporate preferences for different health states. Such measures were developed for resource allocation purposes. In *preference-based* measures, health-related quality of life is summarized as a single number along a continuum that usually extends from death (0.0) to full health (1.0), although states worse than death are possible. Preference scores reflect both health status and the value of that health state to the person being assessed. Far more common are *descriptive* measures in which each item of the instrument or measure is weighted equally and frequency of response is used to calculate scores.

Choosing the Appropriate Measure

Selecting a health measure depends on the purpose of the study and essential characteristics of the study population. For cross-sectional studies, measures that are good at discriminating between different subgroups in a population are important. For longitudinal designs (as in the prospective cohort study proposed by the committee in Chapter 5), measures that are responsive to small changes in health status are important. For health status surveys, generic measures may be particularly useful in comparing levels of health with extant norms and reference values. Generic measures can be supplemented with a condition- or diagnosis-specific instrument if study resources allow and there is a reason to focus on a specific condition. Gulf War veterans as a population group might be described with either a generic measure or with measures specific to an identified condition of concern.

The goals of health assessment include (1) differentiating between people who have a better health status and those who have a worse health status; (2) measuring health over time, which involves identifying the extent to which the status of an individual or a group has changed; and (3) making predictions or diagnoses.

For prediction, the sensitivity and specificity of the measure are both important. For example, an index measure that discriminates well between persons with or without a specific disease, such as thyroid disease, might not include fatigue as an item because fatigue is common among people with and without thyroid disease and therefore not a highly specific indicator for thyroid disease. However, a measure good at discriminating among different groups or individuals may not be right for measuring changes in health status over time. For measuring improvement achieved with a treatment for thyroid disease, level of fatigue would be an essential item because of its importance in the daily lives of people with thyroid disease.

Use of Self-Reported Data

An important consideration in assessing the health of Gulf War veterans is the appropriateness of using self-reported information. Because many of the symptoms and subjective complaints of Gulf War veterans cannot be measured solely with physiological or physical tests, or observable markers such as death, the committee views self-report measures as essential to complement evidence from physiological measurement, physical examination, and performance-based measurements such as treadmill or timed walking tests.

There are strong arguments for taking the individual or patient's point of view into account when assessing health-related quality of life. First, outcomes of care, and therefore prior health status, must be defined broadly enough to include variables important to patients and consumers (IOM, 1990). Second, only patients can describe or relate the symptoms or health states they experience; examples include pain, energy or fatigue, or feelings of sadness. Third, discrepancies are frequently noted between physician and patient judgments (Hall et al., 1976; Jachuck et al., 1982; Sugarbaker et al., 1982; Thomas and Lyttle, 1980). Discrepancies are not surprising because concerns of clinicians and patients often differ. Patients may focus more on felt distress, fatigue, or ability to function whereas physicians may focus more on risk factors, probability of changing health, or specific treatments.

Self-reports of health, like much subjective information, are sometimes perceived as too soft a basis for drawing definitive clinical, research, or policy conclusions. Some observers believe that physiological data, physician observation, and records are inherently more accurate, reproducible, and hard. Several studies have suggested, however, that items of medical history or questionnaire responses obtained through the reports of the individual patient or study participant can be more reproducible than a physician's examination or interpretations of tests (Deyo et al., 1985; Koran, 1975; Pecoraro et al., 1979; Wood et al., 1979). Feinstein (1977) has suggested that patient-generated data may be as reproducible as physiological measures and thus equally as hard. Moreover, medical intervention may improve functional health and quality of life without evidence of physiological improvement (Croog et al., 1986; Kaplan et al., 1984; Million et al., 1982). Medical therapy may also result in evidence of physiological improvement without discernable clinical benefit to patients.

The prospective cohort study recommended by this committee relies on collection of self-reported measures. The committee believes this data source is appropriate, reliable, and justifiable for delineating the health of Gulf War veterans. These veterans suffer from symptoms not easily categorized into clear diagnostic entities. Because of the ill-defined nature of their complaints, much of their health assessment relies upon the descriptions of their complaints to a health-care worker. Indeed, the medical model for assessing and caring for affected individuals itself relies on self-reports.

As demonstrated in the Registry programs of VA and DoD, extensive clinical evaluations of Gulf War veterans have demonstrated that physiological and performance-based measures do not correlate well with the nature and severity of symptoms of Gulf War veterans. Alternative mechanisms to capture the health of Gulf War veterans lack the feasibility of this survey methodology. The committee believes that the data contained in completed surveys of health status will form the basis for hypothesis generation and subsequent in-depth studies of targeted subgroups or promising therapeutic interventions.

Data Collection Instruments

Numerous books describe the process for developing reliable, valid instruments that measure health-related quality of life. Because that process is both time-consuming and expensive, it is best, whenever possible, to use previously tested and validated instruments. Chapter 5 briefly discusses issues concerning construction of new measures or adaptation or modification of existing questionnaires.

Many health status measures are derived from data obtained through surveys and interviews. Traditional survey and interviewing techniques derive from a philosophical, statistical, and practical base that assumes, among other things, instruments of fixed length and format developed through a standardized conceptual and analytic process. Such a "classic" process includes:

- developing a formal conceptual and measurement model;
- developing large "item databases";
- carrying out various quantitative analyses to place items into appropriate categories, "factors," or scales;
- conducting other analyses to reduce items within categories and overall to a manageable number; and
- testing the reliability and validity of the resulting instrument.

The proliferation of instruments in the health-related quality of life literature illustrates this general approach to instrumentation. Such instruments can be applied individually or in groups to obtain responses. The field of psychometrics has adapted and validated the psychophysical methods for use in fields that lack objective physical measures for assessing health. Studies demonstrate that people can accurately judge loudness of sound, intensity of an electric shock, or the brightness of light (McDowell and Newell, 1996).

In recent years a new approach has arisen that goes by various names, such as item-response theory (IRT), computer-adaptive testing (CAT), or dynamic testing. These approaches attempt to use a very wide universe of questions, not to dictate a fixed format or length to the instrument, and to employ statistical calculations to reduce the necessary number of questions, per individual respondent, to as few as possible consistent with preestablished reliability thresholds.

As a practical matter, therefore, IRT or CAT approaches rely heavily on computer technologies. The advantages of these types of approaches lie in the presumptively greater efficiency of the questionnaire for large groups of subjects and the broader set of concepts, constructs, topics of interest, scales, and items from which to draw. Applied originally in the field of educational testing (for instance, in licensing examinations for health professionals), IRT or CAT techniques are taking hold in the health status assessment arena as well.

SUMMARY

This chapter has briefly reviewed the concepts this committee is using to characterize its view of health, as it should be measured in addressing the four questions set out in Chapter 1. The committee has identified five major concepts—death and duration of life, impairments, functional status, general health perceptions, and opportunity—as the core elements to be studied. The next two chapters discuss the means by which the committee believes these aspects of the health of Gulf War veterans and identified comparison groups should be investigated over time. Chapter 4 presents the committee's proposal for a "research portfolio" for studies of the health of Gulf War veterans. Chapter 5 outlines the foundation of that research portfolio, a prospective cohort study called the Gulf War Veterans Health Study (GWVHS).

4

The Gulf War Veterans Health Research Portfolio

As discussed in Chapter 1, the first of the three components in the charge to the IOM committee called for identifying questions important in evaluating the health and well-being of active-duty troops and veterans who were deployed to the Gulf War. Chapters 2 and 5 address the second component of the charge: to identify issues to be addressed in the development of study designs and methods that would be used to answer such questions. The third component of the charge to the committee was to develop a research design (or designs) that could be used to address such questions. Because these questions about the health of Gulf War veterans are diverse, efforts to address them in a thorough manner will require the application of various types of research and health measurement (e.g., population monitoring, treatment effectiveness, and clinical practice quality).

RESEARCH PORTFOLIO TO GUIDE STUDIES OF THE HEALTH OF GULF WAR VETERANS

In the committee's judgment, a single study cannot satisfy all information needs about the health of Gulf War veterans (and the various comparison groups noted in the questions in Chapter 1). The committee also recognizes that many completed studies have already made important contributions to our understanding of the problems affecting the health of Gulf War veterans and that other valuable studies are under way or will be undertaken. Various agenda-setting bodies are directing the flow of resources to these investigations. The committee believes, however, that the contributions of future individual studies will be enhanced by a mechanism to coordinate and link these studies. Others, such as the

Presidential Advisory Committee (PAC) (1996b), have also recognized the need for better coordination and oversight of the numerous studies.

Thus, the committee has responded to the task of developing a research design (or designs) that could be used to address questions regarding the health of Gulf War veterans by conceptualizing a "portfolio" of research activities. This portfolio encompasses an array of studies that could be conducted on a variety of topics, using study designs and population subgroups appropriate to the specific questions under investigation. An essential feature of the research portfolio is facilitating linkages across individual studies through the collection of a core set of key data elements, thereby allowing comparisons across all research. This idea is consistent with the observation of the PAC (1996b:31) that "when specific questions from different studies are aimed at obtaining the same information, then consistency [of questions] offers the advantage of allowing future inter-study comparisons."

The committee believes that this portfolio approach will, if implemented, provide a more effective basis for assembling the information needed to achieve a greater understanding of the longer-term health effects of service in the Gulf War. Moreover, the committee views the conceptual framework and the practical, methodological features inherent in the proposed research portfolio as a model that could be used by the Department of Veterans Affairs (VA) and the Department of Defense (DoD) to conduct similar studies of the health consequences of deployment to other conflicts (e.g., Somalia, Bosnia). In fact, the committee believes that inclusion of veterans of such conflicts in some of the studies conducted within the proposed research portfolio will provide a way of distinguishing between health consequences that are unique to the Gulf War and those that emerge from participation in any conflict.

Specifically, **the committee recommends that multiple studies be initiated through a research portfolio with three components: population studies, health services research studies, and biomedical and clinical investigations.**

Further, **the committee recommends that a core set of data on health be collected in all studies and include measures of**

- **death and duration of life,**
- **impairment,**
- **functional status,**
- **health perceptions, and**
- **opportunity.**

These categories of measures reflect the five core concepts of health-related quality of life outlined in Chapter 3.

Research Approaches in the Portfolio

The committee envisions the research portfolio as providing a basis for developing the research agenda—that is, to plan, implement, and coordinate the variety of studies—needed to address diverse veterans' health issues. The portfolio encompasses three principal categories of research: population studies, health services research, and clinical and biomedical investigations. Figure 4-1 illustrates the research portfolio, with the horizontal bars representing the three major categories of research used to study the health of Gulf War veterans.

Population studies are conducted to measure and track levels and trends in health status and health outcomes, risk factors, the use of health services, and other health correlates (as described in Chapter 3) for entire populations, such as communities, or in the context of this report, the population of Gulf War veterans (and specific comparison groups). Such studies are also used to investigate the course of disease and illness. Data are often collected through surveys of statistically valid samples of the population.

Health services research is a multidisciplinary field that investigates the structure, processes, and effects of health care services (IOM, 1995). It may involve health status assessment to establish priorities, examine the effectiveness of health policies and programs, and allocate resources, using a variety of data sources. Within this category, the committee would include program evaluation, policy analyses, and health status assessments to select treatments and monitor individual patient outcomes (often typically with health data obtained from medical examination and testing).

Clinical and biomedical investigations are used to test the efficacy of diagnostic and therapeutic interventions. They are also used to examine the etiology of diseases and less well understood health problems such as those reported by Gulf War veterans.

In developing a research design (or designs) that could be used to address questions regarding the health of Gulf War veterans, the Committee decided to focus on the population studies level of the research portfolio. This conclusion was based on several considerations.

First, it is necessary to determine to extent to which the Gulf War veteran population experiences health problems. Such an overview is best obtained through a population monitoring study. Second, it is necessary to determine whether identified problems are unique to Gulf War veterans or are shared by other populations and, therefore, not specifically related to service in the Gulf War. Third, it is important to obtain information that will generate hypotheses that can guide more detailed studies at the two other levels of the research portfolio. In Chapter 5, the Committee describes in more detail a prospective cohort study—the Gulf War Veterans Health Study—that is intended to provide the population-based foundation for the research portfolio.

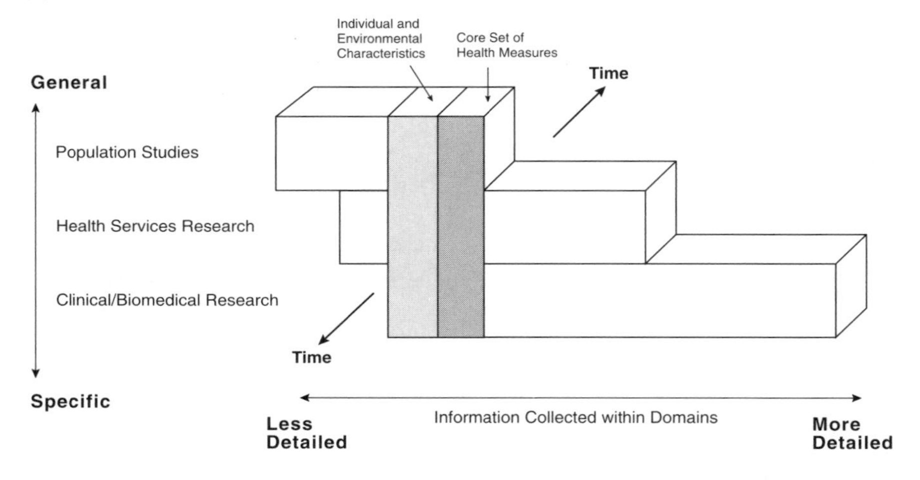

FIGURE 4-1 Research portfolio model for studies of the health of Gulf War veterans.

Dimensions of the Research Portfolio

Figure 4-1 illustrates three dimensions that characterize the range of studies that the committee sees encompassed by the research portfolio: scope of the studies in terms of coverage of the domains of health, level of detail at which studies address the domains of health and health correlates, and the time frame for studies.

Scope of Studies

Study scope is shown as ranging from general to specific on the vertical axis. Population studies fall at the general end of this general-to-specific axis. The prospective cohort study, described in Chapter 5, will cover the five domains of health discussed previously, that is, death and duration of life, impairment, functional status, health perceptions, and opportunity. From the prospective cohort study might emerge an epidemiological study to clarify the incidence, prevalence, and determinants of cognitive impairments among Gulf War veterans and comparison groups.

At the other end of this axis are clinical and biomedical investigations. These are studies in which health data may be the most person-specific because the studies are focused on more specific health or illness questions, often investigated, for instance, in clinical research of biological markers or randomized clinical trials (RCT). As one case in point: results from the prospective cohort study might indicate that fatigue is more prevalent among Gulf War veterans than among comparison populations. Thus, for patients with severe fatigue problems, an RCT of physical therapy and conditioning might be given high priority. Another example is trials of new pharmacotherapeutic agents for use among patients with serious joint pain.

At an intermediate point in terms of scope are health services research studies. Such studies might address a wide variety of topics. With respect to quality of care provided to Gulf War veterans and other comparison groups for specific ailments or problems, a study could examine whether the process and/or outcomes of care differ among these groups. Similarly, one might investigate the cost-effectiveness of alternative interventions for managing Gulf War veterans and other patient populations with chronic fatigue syndrome. In addition, issues of timely access to appropriate, necessary care could warrant investigation; for instance, one might research whether veterans who participate in the VA or DoD Gulf War registries report better access to specialty care.

Level of Detail on Domains of Health and Health Correlates

The horizontal axis of Figure 4-1 represents the level of detail at which studies would address various domains of health and quality of life. For example, a population monitoring study that explores the prevalence and correlates of disability among Gulf War veterans may ask about limitation in physical activity (e.g., walking and performing basic activities of daily living), as well as a large number of additional domains. However, a clinical study of patients with a diagnosis of rheumatoid arthritis is likely to collect detailed information about both upper and lower body function but only core information about other domains such as cognitive function.

Time

The third axis in this model, as shown in Figure 4-1, represents the time dimension of studies in the research portfolio. It is important to note that the length of time it takes to complete a study varies with the type of study conducted, therefore the time axis will show variation in depth. Among population studies, for example, a longitudinal cohort study with repeated administrations of the survey instrument will require a much longer time period than will a single cross-sectional study based on a one-time survey. Including time in the portfolio model also takes into account the possibility of selecting measurement strategies that incorporate time as part of their conceptual framework, such as in the quality-adjusted life year approach to health status and quality-of-life measurement.

Core Data Sets

The core set of measures for all three research approaches should include both measures of health, and likely correlates of health, as discussed in Chapter

3. For comparisons between studies, each of these characteristics should be assessed using the same data collection instruments and methods. Individual investigators could choose to measure any of these characteristics in more detail, as fits the purpose of their studies. In the committee's view, adopting a common minimum set of measures will overcome some of the problems of interpreting Gulf War research findings.

Figure 4-1 illustrates the inclusion of core data sets in research portfolio studies with two shaded vertical columns. The first column represents data on individual and environmental characteristics, for instance, protective and dangerous environmental exposures. This also includes information on personal characteristics such as age, gender, and education, as well as selected information about occupation, socioeconomic status, and military service. The second column represents the core set of health measures. Conceptually, identical data within these two columns would be collected in any investigation carried out within any of the portfolio's three research approaches.

Thus, in addition to the core set of data on health identified in the previous recommendation in this chapter, **the committee recommends that a core set of data on the correlates of health be collected in all studies. These data should include measures of individual and environmental characteristics that are associated with differences in health. Individual characteristics of interest include**

- **biology and life course,**
- **lifestyle and health behavior,**
- **illness behavior,**
- **personality and motivation, and**
- **values and preferences.**

Environmental characteristics of interest include

- **social and cultural,**
- **economic and political,**
- **physical and geographic, and**
- **health and social care.**

For population studies (the top bar in Figure 4-1), the shaded vertical columns at the right-hand side of the bar show the core set of measures on health-related quality of life as being one of the more detailed sets of items. For clinical and biomedical investigations (the bottom bar), the core columns fall at the left-hand side of the horizontal bar, indicating that these measures are a small set of the totality of constituent domains included in clinical investigations. The two columns appearing in the health services research row are an intermediate position in the data collection strategy.

To link the studies of the research portfolio to national-level data for the general population, it would be possible to collect in all portfolio studies the same basic data on perceived health and activity limitations that are collected in the National Health Interview Survey (NHIS), an ongoing survey of the health of the general population in the United States. That is, the same basic set of core data could be collected in population monitoring, health services research, and clinical and biomedical investigations.

With this design approach, information on perceived health and activity limitation as well as their combined measure, the Health and Activity Limitation Index, collected in smaller and focused clinical study groups, would be directly comparable with that collected on representative samples of the U.S. population. With Gulf War veterans as the clinical study population, such comparisons allow examination of the veterans' health status relative to that of their sociodemographic peers in the general population. Similarly, data on the health of veterans participating in the clinical investigations could be linked to findings from epidemiologic or health services studies, if the same core health status measures are used in each.

Linking the portfolio of studies to an ongoing national survey such as the NHIS through a core of common measures provides a basis for distinguishing changes in health-related quality of life that are due to changes in health care from those changes that are due to other factors, such as economic conditions experienced by veterans since the War, versus participation in the War. For example, analysis of the Health and Activity Limitation Index over the 1984–1994 interval indicated that the decline in population health was in part due to the economic recession that occurred in the early 1990s (Erickson, in press).

Other Information Beyond the Core

The white space in the horizontal bars of Figure 4-1 illustrates that studies will include information other than that in the core set of measures. The types of data collected will vary according to the purpose of the various investigations, as represented by the horizontal axis. For example, health data collected in population studies are less likely to consist of physiological measures, such as blood pressure readings, than are data collected for use in clinical or biomedical investigations. Similarly, health services research studies are likely to focus more on use of health care services or costs (e.g., drawing on elements of the Medical Expenditure Panel Survey) than will either population studies or clinical investigations.

SUMMARY

To guide and coordinate studies of the health of Gulf War veterans, the committee proposes a research portfolio that will encompass population studies, health services research studies, and clinical and biomedical investigations.

Within this portfolio, high priority should be assigned to conducting a prospective cohort study that can detect differences in health status among specific populations (e.g., Gulf War veterans and nondeployed veterans) and measure changes in health over time. This approach provides the basis for assessing the extent to which Gulf War veterans are experiencing poorer health compared to other relevant veteran and nonveteran populations.

The following chapter describes the proposed design for this prospective cohort study. It also discusses in more detail statistical and other factors that need to be taken into account in mounting such a study.

5

Gulf War Veterans Health Study

The previous chapter presented the committee's recommendation for the development of a research portfolio to guide studies of the health of Gulf War veterans. Within this portfolio, a prospective cohort study is seen as an essential foundation for the broader array of research. We have called this the Gulf War Veterans Health Study (GWVHS).

As of March 1998, the Department of Veterans Affairs (VA) was projecting cumulative expenditures of $115.2 million for research related to Gulf War health issues from FY 1994 through FY 1998 (VA, 1998a). The types and foci of research topics encompassed by these federally funded projects are broad and include studies at all levels of the research portfolio developed by the committee.

The contributions of the GWVHS and associated research portfolio to these myriad federal efforts are twofold. First, although several studies have focused on Gulf War veterans' health, there has been no population-based study that is both representative of the entire U.S. Gulf War veteran population and that measures current health status, compares it to groups of other veterans and the general public, and provides for the longitudinal tracking through time of changes in health status. To understand the extent to which service in the Gulf War may have affected the health of individuals deployed to this conflict, one must implement such a prospective cohort study.

Second, despite the numerous and high-quality research efforts undertaken at all levels of investigation, no mechanism has been in place that allows comparisons of health to be made between the individual study populations on important and key factors. The research portfolio designed by the committee and described in Chapter 4 would accomplish this.

To understand the contributions of the Gulf War, as well as other conflicts, to the long-term health of the men and women who served, the committee concluded that implementing a prospective cohort study that provides for comparisons with other populations will be necessary. Additionally, collecting a common set of data on key factors in all newly funded efforts is important to enable analysts to make essential comparisons and to conduct key analyses of research results.

For these reasons, the committee believes that implementation of the prospective cohort study—the GWVHS—and the accompanying research portfolio merits the time, effort, and cost required. This chapter describes the key features of this study. It does not, however, provide detailed design specifics, because the committee believes that the principal investigator who will actually implement the GWVHS should develop those aspects. Rather, the committee has provided a template for the GWVHS that addresses the questions the study is intended to answer, its general design, sampling and scheduling issues, considerations for the selection of survey modes and instruments, a proposed pilot study, cost information, ethical considerations, and oversight responsibilities.

STUDY QUESTIONS AND DESIGN

The specific design of a study is dictated by the research questions to be addressed and the target population to be studied. As discussed in Chapter 1, a variety of questions are being asked about the health of Gulf War veterans by the veterans themselves and by other interested parties, including Congress, the VA, and the DoD. The committee believes that among these many questions, of fundamental importance is ascertaining how many Gulf War veterans are suffering from health problems that affect their ability to function; whether the nature of those problems and the frequency of their occurrence in the veteran population are consistent with the experience of the general public or other groups of veterans; and whether the health of Gulf War veterans is getting better, staying the same, or deteriorating with the passage of time.

Because these fundamental questions address both the health of Gulf War veterans at specific points in time and changes in their health over time, **the committee recommends that a prospective cohort study of the population of Gulf War veterans be conducted. Such a study should include appropriate comparison groups.**

Additionally, **the committee recommends that the prospective cohort study investigate the following four questions:**

1. **How healthy are Gulf War veterans?**
2. **In what ways does the health of Gulf War veterans change over time?**
3. **Now and in the future, how does the health of Gulf War veterans compare with that of:**

- **the general population;**
- **persons in the military at the time of the Gulf War but not deployed;**

• persons in the military at the time of the Gulf War who were deployed to nonconflict areas; and

• persons in the military deployed to other conflicts, such as Bosnia or Somalia?

4. What individual and environmental characteristics are associated with observed differences in health between Gulf War veterans and comparison groups?

Further, the committee recommends that this prospective study serve as the foundation for a portfolio of needed research activities. (The research portfolio was discussed in Chapter 4.) As noted earlier, the prospective cohort study is designated by the committee as the Gulf War Veterans Health Study (GWVHS). Appendix B provides a more detailed discussion of the design, methodological and implementation issues.

Although the specific decisions regarding data to be collected and modes of administration to be used are left to those actually implementing the GWVHS, **the committee recommends that the prospective cohort study incorporate the following features:**

• multiple cohorts, one for each group of interest;

• multistage sampling with initial cluster sampling followed by stratified random sampling within clusters;

• random and representative selection of participants within clusters with hypothesis-driven oversampling of specific population subgroups; and

• multiple modes of interviewing, including telephone and in-person interviewing.

Use of a Panel Design

In selecting the prospective cohort design the committee evaluated various alternatives and considered several features of the Gulf War veteran population. Based on these considerations, the committee concluded that either a permanent panel design or repeated panel surveys without temporal overlap is the preferred choice. As reflected in the questions the GWVHS is designed to answer, there is interest in both the level of health of Gulf War veterans and changes in their health over time. Panel studies with temporal overlap (rotating panel studies) are the commonly used approach when both levels and change are of interest. Use of this design calls for periodic selection of a new panel from each of the cohorts of interest because the cohorts are continuously gaining and losing members. That is, refreshing the study with a new sample that accurately represents the changed population is usually necessary.

The committee concluded, however, that a study design calling for repeated selection of new panels would not be necessary because the Gulf War veteran population is closed, that is, the population will not gain new members. Membership in this population was determined by participation in the Gulf War, specifically service in the Gulf War theater of operations between August 2, 1990, and June 13, 1991. Because the population of Gulf War Veterans is a closed population, a representative sample can be selected at one point in time; insofar as the sample is successfully followed, it will continue to be representative of the population of Gulf War veterans.

The committee also concluded that the ability to locate Gulf War veterans and members of the comparison groups of veterans is likely to deteriorate over time. Although DoD maintains records with locating information for all individuals who have served in the armed forces, including Gulf War veterans, the accuracy of address information for those who have left military service becomes less reliable as the years pass. The VA's Benefit Identification and Record Locator Subsystem (BIRLS) is another source of locating information for those who are receiving benefits from VA.

As this report is being written, however, 8 years have elapsed since the Gulf War. Given the anticipated deterioration of the address information in DoD and VA records, the committee anticipates that a substantial tracing and tracking effort will be necessary to locate, recruit, and follow a representative sample of Gulf War veterans in the prospective cohort study it recommends. Because the accuracy of information for selecting the initial sample will become less reliable over time, it is important to select a representative sample of Gulf War veterans as soon as possible. To minimize the cost of sample selection, it is desirable to avoid conducting costly recruitment on a periodic basis.

The quality of the study data should be reviewed periodically to determine the extent to which the validity of inference based on the panels is compromised by cumulative attrition or other factors. A cohort study that is well designed and implemented holds the attrition rates over time to the lowest possible level, 5% or less for each wave. If the quality of the panel for each cohort of interest is judged satisfactory, the study would continue following the same panels. It is important, if at all possible, to continue following the same panel throughout the study to allow data collected at baseline to be used to identify predictors and correlates of health status changes at subsequent points in time.

If attrition leads to a serious degradation of the sample, it may be necessary to select a new panel. If the representativeness of the sample is *somewhat* impaired, it may be advisable to implement a hybrid design that continues to follow random subsamples of the initial panels, and also draws new panels for each cohort to make up for the discontinued portion of the initial panels.

Comparison Groups

Key comparison groups are included in the design to provide a basis for identifying unique characteristics of and changes in the health of Gulf War veterans. Comparisons with civilians provide a basis for ascertaining whether levels and trends in the health status of Gulf War veterans reflect the experience of the general population or show differences that are associated with some aspect of military service. Comparisons with those in the military at the time of the Gulf War but not deployed provide a basis for ascertaining whether deployment is associated with differences in levels and trends in health status.

Comparisons with a sample of individuals who were in the military and deployed to a "safe" area provide a basis for ascertaining whether war-theater deployment is associated with health consequences or whether levels and trends in health status of Gulf War veterans are associated with general deployment to the Gulf region. Finally, comparisons with a sample of veterans of other conflicts (e.g., Bosnia and Somalia) provide a basis for ascertaining whether levels and trends in the health status of Gulf War veterans were associated with serving in any conflict or with unique aspects of the Gulf War situation. Other comparison groups could be added, as new conflicts arise in the future, to provide an ongoing basis for assessing the health effects of service in those conflicts for U.S. military personnel.

For the study recommended by the committee, comparison groups are to be sampled and surveyed using the same design as that used for Gulf War veterans to maximize the comparability of the data obtained from all groups. Because members of the Gulf War veteran population and the comparison groups were not randomly assigned to those groups, the comparison will be subject to potential selection bias problems common to all observational studies: Gulf War veterans might be different from the comparison groups even in the absence of the Gulf War experience. To account for such differences, data are to be collected on potential confounding factors, such as sex, years of schooling, and other correlates of health status.

SAMPLING

Sampling Methods

Many issues regarding sample selection procedures must be addressed. The most straightforward approach is to use simple random sampling. A disadvantage of this technique, for purposes of the GWVHS, is that it requires a listing of the entire study population, that is, not only Gulf War veterans but also the comparison groups. Another disadvantage is the potential costliness if data collection involves conducting in-person interviews or evaluations on a study population that is widely dispersed across the country.

Stratified random sampling involves dividing a total population into groups (e.g., based on age, sex, military branch of service), and selecting a random sample within each group. This approach is frequently used when attempting to ensure that certain groups are adequately represented in the sample populations (e.g., women, American Indians). As with simple random sampling, however, every member of the population must be listed and, in addition, must be categorized using the stratification variables.

Cluster sampling is an approach used when lists of an entire population are not available, but when lists of clusters (e.g., schools within a state or school district) are available. A sample of clusters is selected and all members of selected clusters (e.g., all students in a selected school) are included in the study population. Multistage sampling, often referred to as two-stage sampling, involves initial selection of clusters, followed by listing and sampling of members within each cluster to produce the final sample (Henry, 1990). Using this approach, a sample of geographical locales is drawn first. A stratified random sample of individuals is then drawn from each sampled locale.

Geographical clustering of the study population within the sampling units reduces the cost of data collection, particularly for some types of data and modes of survey implementation. For example, some of the analyses included in the GWVHS or the portfolio of studies might require the use of contextual data, such as the supply of health care services in the geographical locale in which the respondent resides. Although some contextual data can be obtained with little effort for all geographical locales, some detailed contextual data require direct data collection in the specific locales being studied. Additionally, if face-to-face interviewing is used to administer questionnaires, a geographically clustered sample is more efficient than random distribution of the study population across the country. Similarly, conducting physical examinations on a subsample of the GWVHS participants would be facilitated by geographical clustering of the sample.

Over time the level of clustering can be expected to dissipate gradually, due to migration. Some of the participants who migrate out of the original locales from which they were recruited will move to a new locale that is part of or near another cluster sampled, making it relatively easy to continue face-to-face interviews. Those who move away from any cluster included in the study are likely to be costly to follow if face-to-face interviews are required.

Nevertheless, the use of geographic clustering is an efficient, cost-effective approach to sampling. The committee believes that such an approach is the most appropriate method for choosing the samples in the GWVHS. Within clusters the samples would be randomly selected and stratified by demographic characteristics such as age, sex, race/ethnicity, and military characteristics such as branch of service and duty status.

Sample Size

Sufficient sample sizes for each cohort in the study are crucial to ensure adequate statistical power to find differences as well as to reliably identify the lack of differences between groups. The sample size required depends, in part, on the baseline level of the variable of interest in the comparison group and the magnitude of difference expected in the affected group. The committee concluded that *final* decisions on necessary sample sizes for the GWVHS should be left to the researchers who will actually implement the study.

To calculate the appropriate sample size, the number of subjects per group must be multiplied by the number of groups used for comparison. As many as five possible comparison groups have been identified for the GWVHS. If it is anticipated that 25% of potential respondents are expected not to respond, the sample must be increased by 25% to achieve the required size. If the sample is to be followed prospectively, the researchers must anticipate some attrition and they may need to choose a reserve cohort.

The following are examples to illustrate possible sample sizes needed for studies comparing Gulf War veterans with other groups. Sample size calculations are made on the basis of formulas used in EpiInfo Version 6 (1/97).

Assume, for example, that one or more poorly defined symptoms were present in about 1% of non-Gulf War veterans and 2% of Gulf War veterans. To establish that this difference is true with 80% power and 5% statistical significance would require a sample of 2,514 veterans in each of the two comparison groups. If a higher level of significance (1%) is desired, a sample of 3,647 veterans in each group is needed.

If the condition under investigation occurs less frequently than in 1% of the population of interest and all other conditions remain similar, a larger sample will be needed. For example, assume that posttraumatic stress disorder is found in 0.7% of non-Gulf War veterans. If it is twice as common among Gulf War veterans, a sample of 3,600 from each group is needed. If the significance level is raised to 1%, a sample of 5,233 in each group is needed.

If the difference in the frequency of the condition is greater than twofold, the necessary sample size is reduced. In the above example, moving from a twofold difference (0.7 to 1.4%) to a threefold difference (0.7 to 2.1%), sample size for 5% significance drops from 3,600 to 1,243; for 1% it drops from 5,233 to 1,784.

Detection of differences in more common conditions requires smaller sample sizes. Assuming major depression occurs in 2.7% of non-Gulf War veterans and 5.4% of Gulf War veterans, a sample size of 908 is needed for each group for a 5% significance level; for significance of 1%, the sample size is 1,317 for each group.

The committee emphasizes that these examples provide the number of individuals that are needed at the end of a study. To allow for anticipated attrition over the course of a longitudinal study, much larger numbers of participants will need to be recruited at the start of the study to ensure adequate sample size at the end. If subgroup analysis is to be performed, larger samples are also necessary. For example, if comparisons are to be made not just between Gulf War veterans and non-Gulf War veterans, but also between male and female veterans of each group, the appropriate sample size will be larger than that required when comparing only Gulf War and non-Gulf War veterans overall. If a cluster-based sample design is used, the power analyses must take into account the design effect due to intracluster correlation, which usually increases the sample size required.

SCHEDULING

An important parameter in the design of longitudinal cohort or panel studies is the frequency with which the participants are surveyed. The committee has designed this study to include a baseline survey with two follow-up surveys (or waves) over a 10-year study period. Ensuring the timely availability of information has been considered in developing the implementation schedule because the information obtained is of value to the Gulf War veterans only during their lifetimes. The committee concluded that the GWVHS should be designed with more frequent data collection in the early years when the information obtained has a longer "useful life" to the Gulf War veterans and, if the study is continued beyond the first 10 years, with less frequent data collection in later years when the information has a shorter useful life.

For the initial study period, the interval between surveys should not exceed 3 years. This interval should allow time for preliminary analysis of results and any survey modifications deemed necessary after a review of survey findings, and it should provide a short enough interval to maximize participant retention. Additionally, each wave of data collection should be conducted throughout the year to avoid the effects of seasonal variation, but there should be no temporal overlap in data collection across waves. Figure 5-1 illustrates the timing suggested for the GWVHS.

To improve retention among the study participants an effort should be made to maintain contact with them during the intervals between studies. Such activities are generally referred to as tracking. During the baseline survey, the interviewer collects contact information from the participants that will help locate them for follow-up surveys. The contact information is usually updated during each follow-up survey. Additional tracking mechanisms that can be used between surveys include sending postcards, birthday cards, and newsletters to the participants at regular intervals; requesting postal notification of change of address; and requesting that participants submit change-of-address information to the study. Incentives can be offered to encourage the participants to provide this information.

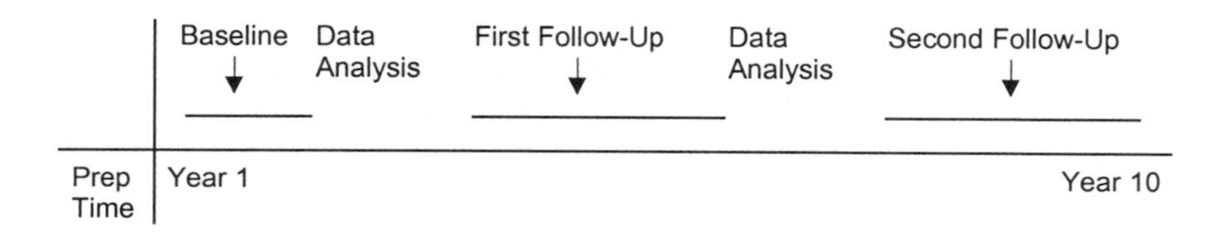

FIGURE 5-1 Timing of the Gulf War Veterans Health Study.

MODE OF SURVEY

The committee believes that those who implement the GWVHS must evaluate multiple modes of data collection based on the types of information to be collected. With the revolution in communication and information technologies in recent years, many data collection methods can be considered, including face-to-face interviews, telephone interviews, distribution and return of questionnaires by mail, or combinations of these modalities.

Computer-assisted interviewing is commonly used in face-to-face and telephone interviews. It can help guide interviewers through the proper sequence of questions, check for invalid responses, and allow responses to be recorded electronically. Audio-assisted interviewing is sometimes used in the face-to-face modality for sensitive topics. It allows respondents to listen and respond to questions privately without direct interaction with the interviewer or household members who might censor responses. Internet-based interviewing via e-mail or the World Wide Web may become a possibility.

Face-to-face interviewing is usually considered the most reliable method, because it allows for the use of auxiliary material such as printed response categories and interviewer prompts that are difficult to use in telephone interviews, and it allows for direct examination of the respondents, such as taking physical measurements. Also, a face-to-face interviewer can ensure that the study participant, rather than a proxy or surrogate, actually answers the questionnaire. However, some evidence suggests that this format may result in more socially desirable responses, such as underreporting of stigmatized behavior, and it is the most costly form of interviewing.

Telephone interviewing is less costly than face-to-face interviews, but it is limited to those who have access to a telephone. Mail surveys are usually the least costly, but they also usually result in a higher nonresponse rate. These survey modalities can encourage respondents to provide honest answers to questions on sensitive topics, but both methods also make it easier for a proxy or surrogate to respond. Regardless of the method of survey administration adopted, GWVHS researchers should allow for proxy assessment of the health status of panel members who die during the study period or who cannot respond directly due to physical, cognitive, or psychological impairments.

IMPROVING RESPONSE RATES

The willingness of Gulf War veterans and members of the recommended comparison groups to participate in the GWVHS will be crucial to the study's success. Because of the longitudinal nature of the study, the demands on participants will be even greater than for participants in a one-time cross-sectional study. In various survey-based studies of Gulf War veterans conducted to date, response rates have ranged from a low of 31% in the study of active duty and reserve personnel based in Hawaii and Pennsylvania, conducted by Stretch et al.

(1995, 1996), to 97% of those located in a survey of women who served in the U.S. Air Force during the Gulf War, conducted by Pierce (1997). Studies that included non-Gulf War comparison groups have generally found lower response rates in those comparison groups (e.g., Goss Gilroy, 1998; Holmes et al., 1998; Iowa Persian Gulf Study Group, 1997; Unwin et al., 1999). Table 5-1 presents response rate information for several such studies conducted.

Maximizing response is an important issue in all survey studies, since almost all such studies fail to obtain participation from some individuals in the sample. Reasons for such nonresponse will vary: some subjects cannot be located or reached, some are too sick to be interviewed, some refuse to be interviewed. If the response rate is low, for example, less than 70%, there is the potential that nonresponse bias may seriously undermine the ability to draw conclusions from the data. Statistical and econometric techniques can reduce impact of nonresponse, but efforts can and should be made to maximize response rates. (A detailed discussion of these issues can be found in Appendix B, section 5 entitled Nonresponse, Attrition, Tracking, and Tracing.)

The committee identified two particular areas that should be explored as approaches for improving participation in the GWVHS: veteran participation in organizing and implementing the study, and use of incentives. The "Iowa Study" (Iowa Persian Gulf Study Group, 1997) illustrates the potential benefit of veteran participation. Principal investigators identified their Public Advisory Committee, composed of members from veterans organizations, as a key factor in achieving high participation. This committee provided input from the beginning stages of the study, and assisted in generating participation of the veteran community. This IOM committee believes that veteran participation in organizing and implementing the GWVHS is a key element in facilitating the participation of veterans, not only from the Gulf War cohort, but also from other veteran comparison groups.

Another approach to improving response rates is the use of incentives. Research has shown that monetary incentives are effective at increasing response rates; are more effective when prepaid than when promised; may induce greater respondent commitment and, therefore, improve quality of data obtained; are most effective at increasing the response rates of individuals with lower income and less education; when combined with follow-up mailings, are significantly more effective than one, two, or three follow-up mailings without incentives; and added costs are likely to be offset by savings in the costs of follow-up activities (Armstrong, 1975; Berk et al., 1987; Berlin et al., 1992; Cannell and Henson, 1974; Duffer et al., 1996; James and Bostein, 1990; Kulka, 1993; Linsky, 1975; and O'Keefe and Homer, 1987). Because of the importance of achieving and maintaining representative participation in the study, the committee urges the principal investigators implementing the GWVHS to consider the use of monetary incentives.

TABLE 5-1 Response Rates in Previous Survey-Based Studies of Gulf War Veterans

Response Rate	Study Characteristics	Author/Citation
97% (of 525 located veterans) 92% at follow-up 1 87% at follow-up 2	Mail survey; two subsequent survey rounds Study population: women in U.S. Air Force (active duty, reserve, or National Guard) during Gulf War period	Pierce, 1997
95% (of 620)	Questionnaires/assessments completed during weekend training sessions (psychological symptoms) Study population: Gulf War veterans from reserve units in western Pennsylvania; includes personnel not deployed or deployed to Europe	Perconte et al., 1993
78.4% (of 2,949)	Study population: Ft. Devens ODS Reunion Survey; Army Active Reserve and National Guard soldiers assessed within 5 days of return to United States	Wolfe et al., 1998
Gulf War veterans: 78.3% (of 2,421) Non-Gulf War veterans: 73.0% (of 2,465)	Telephone survey Study population: Active duty, reserve, National Guard troops serving in or during Gulf War period; Iowa as home of record	Iowa Persian Gulf Study Group, 1997
74.4% (of 160) at 1 month 52.5% at 6 months 38.8% at 2 years	Self-administered questionnaires completed at training sessions at 1 month, 6 months, and 2 years following return	Southwick et al., 1993, 1995

Continued

TABLE 5-1 *Continued*

Response Rate	Study Characteristics	Author/Citation
	Study population: returning members of two National Guard reserve units (medical company, military police company)	Southwick et al., 1993, 1995 (*continued*)
74.2% (of 136)	Questionnaires distributed to unit members Study population: single National Guard unit	Sostek et al., 1996
Gulf War veterans: 73.0% (of 4,262) Not deployed: 60.3% (of 5,699)	Mail-out, mail-back survey; one follow-up questionnaire mailing Study population: Canadian Gulf War and Gulf War-era veterans	Health Study of Canadian Forces Personnel (Goss Gilroy, 1998)
Gulf War veterans: 70.4% (of 4,246) Bosnia veterans: 61.9% (of 4,250) Not deployed: 62.9% (of 4,248)	Mail-out, mail-back survey; two follow-up questionnaire mailings Study population: U.K. Gulf War veterans, Bosnia veterans, Gulf War-era veterans	Unwin et al., 1999
Phase I: 52.8% (of 30,000) Phase II: 69.8% (of 30,000)	Phase I: mail-out, mail-back survey; three follow-up mailings Phase II: telephone interviews with nonrespondents Study population: 15,000 U.S. Gulf War veterans, 15,000 Gulf War-era veterans	VA National Health Survey of Gulf War-era veterans and their families (study ongoing)
Index unit: 62% (of 1,083) Unit A: 35% (of 1,520) Unit B: 73% (of 1,141) Unit C: 70% (of 2,407)	Questionnaires distributed during unit training weekends; for Unit C, during 10-day period	Fukuda et al., 1998

Continued

TABLE 5-1 *Continued*

Response Rate	Study Characteristics	Author/Citation
	Study population: Index unit: Air National Guard, Lebanon, Pennsylvania; Unit A: other Pennsylvania Air National Guard, Unit B: Air Force Reserve; Unit C: active Air Force (Florida)	Fukuda et al., 1998 (*continued*)
Deployed: 57.3% (of 517) Nondeployed: 42.2% (of 497)	Mail-out, mail-back survey Study population: activated Air National Guard Unit	Holmes et al., 1998
41% (of 606)	Mail and telephone recruiting; questionnaires completed at one of five sessions Study population: 24th Reserve Naval Mobile Construction Battalion	Haley, et al., 1997
31% (of 16,167)	Questionnaires distributed through units Study population: Active duty, National Guard, and reserve personnel from units in Hawaii and Pennsylvania	Stretch et al., 1995, 1996

SOURCE: Compiled from published articles cited.

PILOT STUDY

Because the GWVHS recommended by this committee requires a major commitment of resources, a pilot study will be essential to determine its feasibility and its cost.

The committee recommends a pilot study be conducted to determine the feasibility and cost of the prospective cohort study. The pilot study should include an assessment of the following points:

- **for each of the five cohorts, identification of the universe from which the sample is to be drawn, especially the Gulf War veteran sample;**

- **willingness of members of each cohort to participate in the baseline study;**
- **modes of data collection; and**
- **use of incentives to maximize response rates.**

DATA COLLECTION INSTRUMENTS

Three main approaches can be followed in identifying instruments to use in measuring selected health domains: use existing, validated instruments in their entirety; modify existing questionnaires (including combining existing questionnaires into larger batteries, which might be construed as a "new" instrument); and construct wholly new instruments.

Using Existing Instruments

Generally, the committee advises that, wherever possible, proposed surveys and studies should attempt to use existing instruments, where such instruments have been documented (ideally in the peer-reviewed literature or through other authoritative avenues) to be reliable, valid, reasonably responsive, and reasonably practical to implement. There are several reasons for this approach. The first reason is resources. Existing validated instruments need not be put through further extensive psychometric testing. Moreover, existing instruments offer the potential advantage that population norm data are available, providing a basis for valuable cross-sectional or longitudinal comparisons (e.g., with civilians or with populations facing different types of stressful events).

The second reason is timing. Existing instruments could be put into the field months, or even years, ahead of questionnaires that are developed *de novo*. The third consideration is comparability. It may well be that at least some instruments (or parts of instruments) have been fielded already within the Gulf War veteran population; to the extent they can be used in the proposed studies over time, they will provide a better comparison with previous studies than will newly created instruments. Finally, issues of practicality arise; for example, instruments already used in federally contracted studies have passed the clearance requirements of the Office of Management and Budget and thus are easier to describe and justify.

Identifying instruments is, of course, a critical exercise. Many books review available instruments in terms of their purpose, conceptual basis, reliability and validity, and other characteristics (e.g., Bowling, 1997; Frank-Stromborg and Olsen, 1997; McDowell and Newell, 1996).

Documenting the quality of off-the-shelf instruments that are considered for these purposes will also enhance the perceived and actual credibility of the surveys/studies. GWVHS principal investigators should review published compilations or available databases to ascertain what information is currently available regarding the reliability, validity, and similar properties of existing instruments.

Another approach is direct evaluation of potential instruments against established criteria as discussed by Lohr et al. (1996). A third option is for the designers of these surveys/studies (or VA and DoD) to engage the services of evaluators knowledgeable in the assessment of a wide array of existing instruments for the purposes envisioned for this study, for example, the Medical Outcomes Trust.

Modifying Existing Instruments

Using existing instruments, more or less intact, may blur into modifying existing instruments. This may be desirable, or at least necessary even if not optimal, in several circumstances. For example, combining two or more instruments (e.g., a generic instrument with one or more condition-specific questionnaires) may result in redundant items, and these should be pared such that each item is asked only once. Different instruments are likely to be formatted differently, and achieving a consistent, unique appearance for any "combined" questionnaires to be employed in the GWVHS will necessarily lead to modifications of at least some of the original forms. Such reformatting should be done with care taken not to distort the intent of the original items or response categories.

Length of a survey instrument may prove to be a significant consideration, and here the GWVHS principal investigators may be tempted to pick and choose items from different instruments that may appear to suit their purposes (e.g., adding to one instrument specific items from other instruments). This practice should be discouraged for questions that constitute subsets of a larger questionnaire, unless it is essential in view of interview length or time constraints, as it can make irrelevant the psychometric, scoring, and feasibility properties of the "donating" instrument. Use of such subsets of items from existing instruments requires methodological research on the subset to ensure that it is acceptable to use.

Creating New Instruments

If, in the end, investigators of the proposed studies choose not to use existing instruments at all, then they will face the challenge of constructing a new instrument. By and large, this requires attention to the following types of steps:

- laying out the conceptual model for the domains to be measured;
- developing, in an iterative process, successively smaller pools of items to address the domains laid out in the conceptual model; this work can be done through review of existing instruments, review of the literature, and use of focus groups or other cognitive testing procedures;

• pretesting a draft instrument in the population(s) of interest, with specific attention to reliability (replicability in the test–retest sense), internal consistency (where appropriate), feasibility, and various forms of validity;

• revising the instrument as needed in response to problems identified in the pretesting stage; and

• developing (and testing) different versions of these instruments (e.g., both self- and interviewer-administered models).

When the measurement model calls for instruments for which pure psychometric properties are not considered relevant, then the developers must undertake equivalent testing activities and document the characteristics of the final instrument in other ways. Care should be taken to *format* any new instruments in accordance with accepted principles of format and layout (e.g., those of Jenkins and Dillman, 1977). When instruments are to be administered in a computer-assisted mode, additional efforts are necessary to develop and test the computer programs that are to be used.

Developing and testing data collection instruments can take months and should be led by an experienced instrument development team. Space does not permit detailed description of all the steps of instrument construction, but guidelines for such endeavors can be found in sources such as Fowler (1995).

COST

The committee recognizes that the cost of implementing the recommended prospective cohort study is not insignificant. Recent estimates of the cost of other large national panel surveys offer a range of reference points. The Health and Retirement Study, conducted by the Institute for Social Research of the University of Michigan for the National Institute on Aging, has an estimated cost of $13.4 million for a 2-year cycle. The household component of the Medical Expenditure Panel Survey of the Agency for Health Care Policy and Research has a budget of $13.8 million for six rounds of interviews with a single panel over a 2.5-year study period. The cost of the Panel Study of Income Dynamics, also conducted by the Institute for Social Research and funded by the National Science Foundation and other federal agencies, is estimated at $2 to $3 million per year. The budget for the Survey of Income and Program Participation, conducted by the U.S. Census Bureau, is $31 million for FY 1999. These cost estimates for ongoing survey programs do not include the initial costs to select the sample and develop the survey instruments and procedures.

ETHICAL CONSIDERATIONS

As with any survey or experiment involving human participants, ethical considerations must be addressed in implementing the GWVHS. For the most part, these are governed by the requirements and regulations of the principal investigators' institutional review board. Certain issues are prominent, however, because of the features of the research portfolio this committee is proposing, because of considerations of compensation and benefits raised by the study population, and because of potential work-related impact for those remaining in the military.

The results of the prospective cohort study are viewed as providing information that will generate hypotheses for further research at other levels of the portfolio. Such results could, for example, generate the desire to include a subsample of the study population in additional research efforts, such as determining treatment effectiveness for a particular condition or symptom. How and in what ways can the obligations and requirements for confidentiality of individual responses be balanced with the benefits to be derived from identifying subpopulations for future research? How will determinations be made about the validity of requests for information to conduct additional research? Who should have that decision-making authority?

Additionally, questions will arise about whether the individuals enrolled in the study are bearing a disproportionate burden of intrusiveness into their lives over an extended period (relative to those never approached or enrolled in the study). The issue of initial study participation is presumed to be dealt with through informed consent, but consent for further study participation also should be explicit.

These are examples of some of the important questions that must be addressed prior to implementation of the GWVHS. The following section discusses a mechanism for dealing with these and other important oversight issues relevant to implementing the research portfolio.

INDEPENDENT ADVISORY BOARD

The long-term research strategy envisioned through full implementation of the research portfolio laid out by this committee will involve many participants: individuals from the armed services and the civilian sector, and agencies and organizations from both the public and private sectors. Moreover, many factors in the coming years may exert substantial influence on the particular questions to be asked and the specific types of projects to be carried out within this research strategy. The structure and the assumptions that underlie the committee's proposed study design are dynamic, not static. This is particularly true if the portfolio is applied to investigations of the health effects of contemporaneous conflicts (e.g., Somalia, Bosnia) and to those of as yet unforeseen conflicts, peacekeeping efforts, or wars as well as to the health problems of Gulf War veterans.

Health-related factors likely to be of importance include well-known and predictable phenomena. Aging of the Gulf War population, for example, will bring more of the "traditional" chronic diseases as an overlay on health problems that some veterans now manifest or could experience in the future. At the same time, newly emerging and reemerging diseases, particularly infectious diseases, may complicate the epidemiological picture for veterans. In addition, the number and sophistication of new health technologies—diagnostic, therapeutic, rehabilitative, and preventive—will continue to increase. Understanding of the genetic or molecular basis of disease will also improve, potentially giving rise to approaches to management of illnesses not thought of or considered possible today.

To ensure high-quality research throughout the program the committee believes an independent advisory board should be established to set policies for and monitor the progress of the prospective cohort study and research portfolio of the health of Gulf War veterans and veterans of other military conflicts. Such an advisory board could include the following tasks among its functions:

- provide scientific and public oversight of research on issues related to the health of Gulf War veterans and veterans of other conflicts;
- establish policies regarding data protection and access, and review and award research grants or contracts;
- ensure integration of new research findings to advance understanding and treatment of war-related illnesses; and
- provide advice to federal agencies on research on the health of veterans.

The advisory board should be an independent body to ensure its scientific integrity and public perception of validity of research results. It should be composed of acknowledged leaders in fields pertinent to the investigation of veterans' health issues and should include members who represent the veteran population.

The benefits of such an advisory board are several. First, it provides for a broad range of expertise to participate in the oversight of this major, and complicated, effort to monitor the health of veterans of military conflict. Second, its agenda can be quite broad and thus encompass more than might be accomplished by any single federal department; it could include relatively immediate concerns such as funding priorities as well as longer-term issues such as how to take account of changing disease epidemiology, new health care technologies, or new populations of veterans of other conflicts.

Third, it provides a visible mechanism for public accountability. Finally, such an advisory board can command national attention when it speaks or acts; it is thus in a position to call for direct, immediate, and meaningful action on the conclusions and implications of critical findings.

The functions of the advisory board should include a review of the scientific and methodological merit of proposed and ongoing studies in the research portfolio. This review should take into account not only the research activities being supported or carried out within the structure proposed in this report, but also changes in various other programs within the federal government or the private sector. Of particular concern would be modifications to large national surveys (e.g., the National Health Interview Survey, the Medical Expenditures Panel

Survey, or similar periodic surveys of the Department of Health and Human Services or other federal agencies) that capture epidemiological and other data relevant to the health and health care utilization of veterans. Should any major changes be judged necessary, they can be set in motion in an orderly, but timely way so that consistency with earlier studies and methods is maintained.

Therefore, **the committee recommends that an independent advisory board oversee the conduct of the prospective cohort study. The advisory board should**

- **be an independent, scientific, and policy-oriented body composed of experts in clinical medicine, epidemiology, health status and health outcomes assessment; veterans' health issues; health services research; social, behavioral, physical, and biomedical sciences; survey research; statistics; national health databases; and health policy, along with members of the public who represent Gulf War veterans.**
- **review, in a timely manner, requests for proposals developed by the funding agencies to conduct the prospective cohort study recommended by the committee.**
- **evaluate the methodological design of the cohort study.**
- **set the minimum requirements for policies on**
    - **methods for locating and retaining study participants,**
    - **informed consent,**
    - **respondent burden,**
    - **confidentiality and security of data,**
    - **use of incentives, and**
    - **responsibility for reporting identified individual or public health threats.**
- **evaluate the success of the prospective cohort study at the end of the 10-year study period.**
- **submit a report to Congress every 2 years.**

SUMMARY

A prospective cohort study—the Gulf War Veterans Health Study—comparing the health of Gulf War veterans to other veteran and civilian groups provides the opportunity to obtain important information about the current and long-term health effects of military service in conflicts in which the United States engages. Such information will provide the basis for analyzing the extent to which health status is a function of a particular conflict, of participation in conflict in general, or the result of factors common to both military and civilian populations.

Special effort must be made to ensure the integrity of the study process and public acceptance of findings because of the many questions and concerns voiced by both members of Congress and the public about the openness and completeness of federal efforts to investigate the health problems of Gulf War veterans. To this end, the committee believes it essential to establish an independent advisory board composed of scientists and members of the public to oversee the GWVHS process and to report findings.

6

Conclusion

Numerous investigations and research efforts have been undertaken because of concern about the impact of the Gulf War on the health of U.S. troops who served in that conflict. Some of these efforts have addressed the federal government's preparedness to meet its obligations and responsibilities to protect U.S. military service members, veterans, and their families. Others have attempted to determine what health effects might be attributed to service in the Gulf War. Still others have tried to identify possible causes for the myriad reports of health problems among Gulf War veterans.

The continuing focus on the health problems of Gulf War veterans is attributable in no small part to the efforts of individual veterans and the organizations that represent them. These veterans have tirelessly kept before the public and the Congress the idea that more needs to be done to help those veterans who are experiencing health problems they believe are due to their service in the Gulf War.

Although the adequacy of the government's response has been criticized, the Departments of Veterans Affairs (VA), Defense (DoD), and Health and Human Services (HHS) have expended enormous effort and resources in attempts to address a number of important issues related to the health of Gulf War veterans. VA and DoD have implemented clinical diagnostic programs in which more than 100,000 Gulf War veterans have participated. They have funded more than 120 distinct research projects on Gulf War veterans' illnesses, covering the following areas of research:

- prevalence of and risk factors for symptoms and alterations in general health status,
- brain and nervous system function,
- reproductive health,

- immune function,
- mortality experience,
- environmental toxicology,
- chemical weapons,
- depleted uranium,
- pyridostigmine bromide,
- leishmaniasis,
- interactions of exposures,
- prevention of diseases and illnesses,
- diagnosis of conditions related to Gulf War service, and
- treatment of conditions related to Gulf War service.

The findings of these efforts have contributed valuable information to our attempts to understand the causes and consequences of Gulf War veterans' illnesses, yet fundamental questions remain. We do not know the extent to which the population of Gulf War veterans is experiencing health problems that they believe are related to service in the Gulf, nor do we know whether the health status of the Gulf War population is better than, worse than, or the same as that of veterans who were not deployed to the Gulf War. Additionally, there has been no systematic evaluation of whether the health status of these veterans is changing and, if so, how.

The committee has developed and recommends implementation of a research portfolio and prospective cohort study—the Gulf War Veterans' Health Study (GWVHS)—that it believes will address these questions. Key to this portfolio is the linking of individual studies through the collection of a core set of key data elements on health and its correlates. Such linkage will enhance the contributions of future studies by providing a mechanism that allows for comparisons across all research undertaken. The committee believes that the GWVHS and the broad research portfolio will, if implemented, lead to a greater understanding of the longer-term health effects of service in the Gulf War.

The issues surrounding the health of Gulf War veterans are complex, from both a scientific and a policy perspective. Many believe, correctly or not, that attempts to address these issues have been governed by people with conflicting interests. As long as that view persists, resolving these issues will be difficult. The committee recommends establishing an independent advisory board to oversee the implementation of the GWVHS and accompanying research portfolio to assure the public, the veterans, Congress, the scientific community, and others that all efforts to resolve these issues are being conducted according to the highest standards of scientific integrity and public accountability.

In the more than 8 years since the men and women who served in the Gulf War returned home, many veterans have become ill and believe that their health problems are a consequence of participation in the Gulf War. A schism has developed, with ill veterans and their representatives on one side and the federal agencies charged with addressing veterans' health problems on the other. In the interests of all involved, additional coordinated and concerted efforts must be made to bridge this gap. The committee believes that the recommendations in this report will facilitate that process and urges that they be implemented as speedily as possible.

References

Allen, H.M., and Rogers, W.H. 1997. The Consumer Health Plan Value Survey: Round Two. *Health Affairs* 16(4):156–166.

Araneta, M.R., Moore, C.A., Olney, R.S., Edmonds, L.D., Karcher, J.A., McDonough, C., et al. 1997. Goldenhar Syndrome Among Infants Born in Military Hospitals to Gulf War Veterans. *Teratology* 56:244–251.

Armstrong, B.K., White, E., and Saracci, R. 1992. *Principles of Exposure Measurement in Epidemiology*. New York: Oxford University Press.

Armstrong, J.S. 1975. Monetary Incentives in Mail Surveys. *Public Opinion Quarterly* 39:111–116.

Berk, M.L., Mathiowetz, N.A., Ward, E.P., and White, A.A. 1987. The Effect of Prepaid and Promised Incentives: Results of a Controlled Experiment. *Journal of Official Statistics* 3(4):449–457.

Berlin, M., Mohadjer, L., Waksberg, J., Kolstad, A., Kirsch, I., Rock, D., and Yamamoto, K. 1992. An Experiment in Monetary Incentives. *American Statistical Association: 1992 Proceedings of the Survey Research Methods Section*. Alexandria, Va.: American Statistical Association. Pp. 393–398.

Bowling, A. 1997. *Measuring Health: A Review of Quality of Life Measurement Scales,* 2nd ed. Philadelphia: Open University Press.

Butcher, J.N., Williams, C.L., Graham, J.R., Archer, P.R., Tellegen, A., Ben-Porath, Y.S., and Kammer, B. 1992. *MMPI-A:P—Minnesota Multiphasic Personality Inventory*. Minneapolis: University of Minnesota Press.

Butler, R.W., and Satz, T. 1995. *Personality Assessment of Adults and Children*. In Kapland, H.I., and Sadock, B.J., eds., *Textbook of Psychiatry,* 6th ed. Baltimore, Md.: Williams & Wilkins. Pp. 544–562.

Cannell, C.F., Henson, R., and Hybels, J. 1974. Incentives, Motives, and Response Bias. *Annals of Economic and Social Measurement* 3(2):307–317.

Cattell, R.B., Eber, H.W., and Tatsooka, M.M. 1970. *Handbook for the Sixteen Personality Factor Questionnaire* (16PF). Champaign, Ill.: Institute for Personality and Ability Testing.

Cloningen, C.R. 1987. A Systematic Method for Clinical Description and Classification of Personality Variants. *Archives of General Psychiatry* 44:573–581.

Coker, W.J., Bhatt, B.M., Blatchley, N.F., and Graham, J.T. 1999. Clinical Findings for the First 1,000 Gulf War Veterans in the Ministry of Defence's Medical Assessment Programme. *British Medical Journal* 318:290–294.

Converse, J.M., and Presser, S. 1986. *Survey Questions: Handcrafting the Standardized Questionnaire.* Newbury Park, Calif.: Sage.

Cowan, D.N., DeFraites, R.F., Gray, G.C., Goldenbaum, M.B., and Wishik, S.M. 1997. The Risk of Birth Defects Among Children of Persian Gulf War Veterans. *New England Journal of Medicine* 336(23):1650–1656.

Croog, S.H., Levine, S., Testa, M.A., Brown, B., Bulpitt, CJ., Jenkins, CD., et al. 1986. The Effects of Antihypertensive Therapy on the Quality of Life. *New England Journal of Medicine* 314:1657–1664.

Dahlstrom, W.G., Welsch, G., and Dahlstrom, L. 1972. *An MMPI Handbook.* Vol. 1: *Clinical Interpretation.* Minneapolis: University of Minnesota Press.

Day, L. 1989. *Designing and Conducting Health Surveys.* San Francisco, Calif.: Jossey-Bass.

DeMaio, T.J., ed. 1983. *Approaches to Developing Questionnaires* (Statistical Policy Working Paper 10). Springfield, Va:. Statistical Policy Office, Office of Information and Regulatory Affairs, Office of Management and Budget.

Department of Defense. 1998. Presentation at the Institute of Medicine Workshop on Measuring the Health of Gulf War Veterans, May 7, 1998, Washington, D.C.

Department of Veterans Affairs. 1995. *Manual M-10*, Part III. Washington, D.C.: Department of Veterans Affairs.

Department of Veterans Affairs. 1998a. *Annual Report to Congress: Federally Sponsored Research on Persian Gulf Veterans' Illnesses for 1997.* Washington, D.C.: Department of Veterans Affairs.

Department of Veterans Affairs. 1998b. *Consolidation and Combined Analysis of the Databases of the Department of Veterans Affairs Persian Gulf Health Registry and the Department of Defense Comprehensive Clinical Evaluation Program.* Washington, D.C.: Environmental Epidemiology Service, Department of Veterans Affairs.

Department of Veterans Affairs. 1999. *A Review of the Department of Veterans Affairs Revised Persian Gulf Registry and In-Patient Treatment Files.* Washington, D.C.: Environmental Epidemiology Service, Department of Veterans Affairs.

DeVellis, R. 1991. *Scale Development: Theory and Applications. Applied Social Research Methods Series,* Vol. 26. Thousand Oaks, Calif.: Sage.

Deyo, R.A., and Patrick, D.L. 1989. Barriers to the Use of Health Status Measures in Clinical Investigation, Patient Care, and Policy Research. *Medical Care* 27(Suppl.): S254–S268.

Deyo, R.A., McNiesh, I.M., and Cone, R.O. 1985. Observer Variability in the Interpretation of Lumbar Spine Radiographs. *Arthritis and Rheumatism* 28:1066–1070.

Dillman, D. 1977. *Mail and Telephone Data Collection Methods.* New York: Wiley Interscience.

Duffer, A.P., Lessler, J.T., Weeks, M.F., and Mosher, W.D. 1996. *Impact of Incentives and Interviewing Modes: Results from the National Survey of Family Growth Cycle V Pretest.* In Warnecke, R., ed., *Health Survey Research Methods Conference Proceedings* (DHHS Pub. No. [PHS] 96-1013). Washington, D.C.: Department of Health and Human Services.

Erickson, P. In press. *Relationship Between Health and Economic Well-Being in the United States.* Dissertation completed at George Washington University, Columbian School of Arts and Sciences, May 1996.

Evans, R.G., and Stoddart, G.L. 1994. Producing Health, Consuming Health Care. In Evans, R.E., Barer, M.L., and Marmor, T.R., eds., *Why Are Some People Healthy and Others Not? The Determinants of Health of Populations.* New York: Aldine de Gruyter.

Executive Office of the President. 1998. *A National Obligation: Planning for Health Preparedness for and Readjustment of the Military, Veterans, and Their Families After Future Deployments.* Presidential Review Directive 5. Washington, D.C.: Executive Office of the President.

Federal Committee on Statistical Methodology. 1991. *Seminar on Quality in Federal Data. Session 10: Approaches to Developing Questionnaires.* Washington, D.C.: Statistical Policy Office, Office of Information and Regulatory Affairs, Office of Management and Budget.

Feinstein, A.R. 1977. Clinical Biostatistics XLI. Hard Science, Soft Data, and the Challenges of Choosing Clinical Variables in Research. *Clinical Pharmacology and Therapeutics* 22:485–498.

Fowler, F. 1995. *Improving Survey Questions: Design and Evaluation. Applied Social Research Methods Series,* Vol. 38. Thousand Oaks, Calif.: Sage.

Frank-Stromborg, M., and Olsen, S.J., eds. 1997. *Instruments for Clinical Health-Care Research,* 2nd ed. Sudbury, Mass.: Jones and Bartlett.

Fukuda, K., Nisenbaum, R., Stewart, G., Thompson, W.W., Robin, L., et al. 1998. Chronic Multisymptom Illness Affecting Air Force Veterans of the Gulf War. *Journal of the American Medical Association* 280:981–988.

General Accounting Office. 1993. *Operation Desert Storm: Army Not Adequately Prepared to Deal with Depleted Uranium Contamination* (GAO/NSIAD-93-90). Washington, D.C.: U.S. Government Printing Office.

General Accounting Office. 1995. *Operation Desert Storm: Health Concerns of Selected Indiana Persian Gulf War Veterans* (GAO/HEHS-95-102). Washington, D.C.: U.S. Government Printing Office.

General Accounting Office. 1997. *Gulf War Illnesses: Improved Monitoring of Clinical Progress and Reexamination of Research Emphasis Are Needed* (GAO/NSIAD-97-163). Washington D.C.: U.S. Government Printing Office.

General Accounting Office. 1998a. *Gulf War Veterans: Incidence of Tumors Cannot Be Reliably Determined from Available Data* (GAO/NSIAD-98-89). Washington, D.C.: U.S. Government Printing Office.

General Accounting Office. 1998b. *VA Health Care: Better Integration of Services Could Improve Gulf War Veterans' Care* (GAO/HEHS-98-197). Washington, D.C.: U.S. Government Printing Office.

Gold, M.R., Patrick, D.L., Torrance, G.W., Fryback, D.G., Hadorn, D.C., et al. 1996. Identifying and Valuing Outcomes. In: Gold, M.R., Siegel, J.E., Russell, L.B., and Weinstein, M.C., eds., *Cost-Effectiveness in Health and Medicine.* New York: Oxford University Press.

Goss Gilroy, Inc. 1998. *Health Study of Canadian Forces Personnel Involved in the 1991 Conflict in the Persian Gulf.* Vol. I. Ottawa: Goss Gilroy, Inc.

Gray, G.C., Coate, B.D., Anderson, C.M., Kang, H.K., Berg, S.W., et al. 1996. The Postwar Hospitalization Experience of U.S. Veterans of the Persian Gulf War. *New England Journal of Medicine* 335:1505–1513.

Guyatt, G.H., Eagle, D.J., Sackett, B., Willan, A., Griffin, L., et al. 1993a. Measuring Quality of Life in the Frail Elderly. *Journal of Clinical Epidemiology* 46(12):1433–1444.

Guyatt, G.H., Feeney, D.H., and Patrick, D.L. 1993b. Measuring Health-Related Quality of Life. *Annals of Internal Medicine* 118(8):622–629.

Guzman, J., Ponce de Leon, S., Pita Ramirez, L., Castillo Renteria, C., and Perez Pimentel, L. 1993. Change in the Quality of Life as an Indicator of the Clinical Course of Disease: Comparison of Two Indices. *Revista De Investigacion Clinica* 45(5):439–452.

Haley, R.W., and Kurt, T.L. 1997. Self-Reported Exposure to Neurotoxic Chemical Combinations in the Gulf War. *Journal of the American Medical Association* 277:231–237.

Haley, R.W., Horn, J., Roland, P.S., Wilson, W.B., Van Ness, P.C., Bonte, F.J., et al. 1997a. Evaluation of Neurologic Function in Gulf War Veterans: A Blinded Case-Control Study. *Journal of the American Medical Association* 277(3):223–230.

Haley, R.W., Kurt, T.L., and Horn, J. 1997b. Is There a Gulf War Syndrome? Searching for Syndromes by Factor Analysis of Symptoms. *Journal of the American Medical Association* 277:215–222.

Hall, R., Horrocks, J.C., Clamp, S.E., and Dombal, F.T. 1976. Observer Variation in Assessment of Results of Surgery for Peptic Ulceration. *British Medical Journal* 1(6013):814–816.

Henry, G.T. 1990. *Practical Sampling. Applied Social Research Methods Series.* Vol. 21. Newbury, Calif.: Sage.

Holmes, D.T., Tariot, P.N., and Cox, C. 1998. Preliminary Evidence of Psychological Distress Among Reservists in the Persian Gulf War. *Journal of Nervous and Mental Disease* 186(3):166–173.

Hyams, K.C., Wignall, S., and Roswell, R. 1996. War Syndromes and Their Evaluation: From the U.S. Civil War to the Persian Gulf War. *Annals of Internal Medicine* 125:398–405.

Institute of Medicine. 1990. *Medicare: A Strategy for Quality Assurance.* Vols. I and II. Washington, D.C.: National Academy Press.

Institute of Medicine. 1995. *Health Consequences of Service During the Persian Gulf War: Initial Findings and Recommendations for Immediate Action.* Washington, D.C.: National Academy Press.

Institute of Medicine. 1996a. *Evaluation of the U.S. Department of Defense Persian Gulf Comprehensive Clinical Evaluation Program.* Washington, D.C.: National Academy Press.

Institute of Medicine. 1996b. *Health Consequences of Service During the Persian Gulf War: Recommendations for Research and Information Systems.* Washington, D.C.: National Academy Press.

Institute of Medicine. 1997a. *Adequacy of the Comprehensive Clinical Evaluation Program: A Focused Assessment.* Washington, D.C.: National Academy Press.

Institute of Medicine. 1997b. *Adequacy of the Comprehensive Clinical Evaluation Program: Nerve Agents.* Washington, D.C.: National Academy Press.

Institute of Medicine. 1998. *Adequacy of the VA Persian Gulf Registry and Uniform Case Assessment Protocol.* Washington, D.C.: National Academy Press.

Iowa Persian Gulf Study Group. 1997. Self-Reported Illness and Health Status Among Gulf War Veterans: A Population-Based Study. *Journal of the American Medical Association* 277(3):238–245.

Ismail, K., Everitt, B., Blatchley, N., Hull, L., Unwin, C., David, A., and Wessely, S. 1999. Is There a Gulf War Syndrome? *Lancet* 353:179–182.

Jachuck, S.J., Brierley, H., Jachuck, S., and Wilcox, P.M. 1982. The Effect of Hypotensive Drugs on the Quality of Life. *Journal of the Royal College of General Practitioners* 32(235):103–105.

James, J.M., and Bolstein, R. 1990. The Effect of Monetary Incentives and Follow-Up Mailings on the Response Rate and Response Quality in Mail Surveys. *Public Opinion Quarterly* 54:346–361.

Jenkins, C.R., and Dillman, D.A. 1997. Toward a Theory of Self-Administered Questionnaire Design. In: Lyberg, L., Biemer, P., Sollins, M., DeLeeuw, E., Dippo, C., Schwarz, N., and Trewin, D., eds., *Survey Measurement and Process Control*. New York: Wiley.

Kang, H.K., and Bullman, M.S. 1996. Mortality Among U.S. Veterans of the Persian Gulf War. *New England Journal of Medicine* 335:1498–1504.

Kaplan, R.M., Atkins, C.J., and Timms, R. 1984. Validity of a Quality of Well-Being Scale as an Outcome Measure in Chronic Obstructive Pulmonary Disease. *Journal of Chronic Disease* 37:85–95.

Knoke, J.D., and Gray, G.C. 1998. Hospitalization for Unexplained Illnesses Among U.S. Veterans of the Persian Gulf War. *Emerging Infectious Diseases* 4:211–219.

Koran, I.M. 1975. The Reliability of Clinical Methods, Data, and Judgments. *New England Journal of Medicine* 29:642–646, 695–701.

Kroenke, K., Koslowe, P., and Roy, M. 1998. Symptoms in 18,495 Persian Gulf War Veterans: Latency of Onset and Lack of Association with Self-Reported Exposures. *Journal of Occupational and Environmental Medicine* 40:520–528.

Kulka, R.A. 1993. *A Brief Review of the Use of Monetary Incentives in Federal Statistical Surveys.* (Final Report of a Symposium on Providing Incentives to Survey Respondents at the John F. Kennedy School of Government, Cambridge, Mass., October 1–2, 1992.) Washington, D.C.: Council of Professional Associations on Federal Statistics/Office of Management and Budget.

Lessler, J., and Forsyth, B. 1996. *A Coding System for Appraising Questionnaires*. In: Schwartz, and Sudman, N.S., eds., *Answering Questions: Methodology for Determining Cognitive and Communicative Processes in Survey Research.* San Francisco, Calif.: Jossey-Bass. Pp. 259–291.

Lezak, M.D. 1995. *Neuropsychological Assessment*, 3rd ed. New York: Oxford University Press. Pp. 767–790.

Linsky, A.S. 1975. Stimulating Responses to Mailed Questionnaires: A Review. *Public Opinion Quarterly* 39:82–101.

Lohr, K.N., Aaronson, N.K., Alonso, J., Burnham, M.A., Patrick, D.L., Perrin, E.B. and Roberts, J.S. 1996. Evaluating Quality-of-Life and Health Status Instruments: Development of Scientific Review Criteria. *Clinical Therapeutics* 18(5):979–992.

McDowell, I., and Jenkinson, C. 1996. Development Standards for Health Measures. *Journal of Health Services Research Policy* 1(4):238–246.

McDowell, I., and Newell, C. 1996. *Measuring Health: A Guide to Rating Scales and Questionnaires,* 2nd ed. New York: Oxford University Press.

Million, R., Hall, W., Nilsen, K.H., Baker, RD., Jayson, MI., et al. 1982. Assessment of the Progress of the Back-Pain Patient. *Spine* 7:204–212.

Millon, T. 1985. The MCMI Provides a Good Assessment of DSM-III Disorders: The MCMI-II Will Provide Even Better. *Journal of Personality Assessment* 49:379–387.

Morgan, C.A., Klingham, P., Nicolaou, A., and Southwick, S.M. 1998. Anniversary Reactions in Gulf War Veterans: A Naturalistic Inquiry 2 Years After the Gulf War. *Journal of Traumatic Stress* 11:165–171.

Nunnally, J.C., and Berstein, I.H. 1994. *Psychometric Theory*, 3rd ed. New York: McGraw-Hill.

O'Keefe, T.B., and Homer, P.M. 1987. Selecting Cost-Effective Survey Methods: Foot-in-Door and Prepaid Monetary Incentives. *Journal of Business Research* 15:365–376.

Parsons, T. 1958. Definition of Health and Illness in the Light of American Values and Social Structure. In Gartly, J.E., ed., *Patients, Physicians, and Illness: A Sourcebook in Behavioral Science and Health.* New York: Free Press. Pp. 165–187.

Patrick, D.L., and Erickson, P. 1993. *Health Status and Health Policy: Quality of Life in Health Care Evaluation and Resource Allocation*. New York: Oxford University Press.

Payne, S. 1980. *The Art of Asking Questions.* Princeton, N.J.: Princeton University Press.

Pecoraro, R.E., Inui, T.S., Chen, M.S., Plorde, DK., Heller, JL., et al. 1979. Validity and Reliability of a Self-Administered Health History Questionnaire. *Public Health Reports* 94:231–238.

Perconte, S.T., Wilson, A.T., Pontius, E.B., Dietrick, A.L., and Spiro, K.J. 1993. Psychological and War Stress Symptoms Among Deployed and Non-Deployed Reservists Following the Persian Gulf War. *Military Medicine* 158:516–521.

Pierce, P. 1997. Physical and Emotional Health of Gulf War Veteran Women. *Aviation, Space, and Environmental Medicine* 68:317–321.

Presidential Advisory Committee on Gulf War Veteran's Illnesses. 1996a. *Final Report.* Washington, D.C.: U.S. Government Printing Office.

Presidential Advisory Committee on Gulf War Veteran's Illnesses. 1996b. *Interim Report.* Washington, D.C.: U.S. Government Printing Office.

Presidential Advisory Committee on Gulf War Veteran's Illnesses. 1997. *Special Report.* Washington, D.C.: U.S. Government Printing Office.

Proctor, S.P., Heeren, T., White, R.F., Wolfe, J., Borgos, M.S., et al. 1998. Health Status of Persian Gulf War Veterans: Self-Reported Symptoms, Environmental Exposures, and the Effect of Stress. *International Journal of Epidemiology* 27:1000–1010.

Rossi, P.H., Wright, J.D., and Anderson, A.B., eds. 1983. *Handbook of Survey Research.* New York: Academic Press.

Scientific Advisory Committee, Medical Outcomes Trust. 1999. *Evaluating Health Outcome Measures: The Medical Outcomes Trust Approach.* Boston: Medical Outcomes Trust.

Segen, J.C., ed. 1992. *The Dictionary of Modern Medicine.* Park Ridge, N.J.: Parthenon.

Sostek, M.B., Jackson, S., Linevsky, J.K., Schimmel, E.M., and Fincke, B.G. 1996. High Prevalence of Chronic Gastrointestinal Symptoms in a National Guard Unit of Persian Gulf Veterans. *American Journal of Gastroenterology* 91:2494–2497.

Southwick, S.M., Morgan, C.A., Darnell, A., Bremner, D., Nicolaou, A.L., et al. 1995. Trauma-Related Symptoms in Veterans of Operation Desert Storm. *American Journal of Psychiatry* 152:1150–1155.

Southwick, S.M., Morgan, C.A., Nagy, L.M., Bremner, D., Nicolaou, A.L., et al. 1993. Trauma-Related Symptoms in Veterans of Operation Desert Storm: A Preliminary Report. *American Journal of Psychiatry* 150:1524–1528.

Stretch, R.H., Bliese, P.D., Marlowe, D.H., Wright, K.M., Knudson, K.H., and Hoover, C.H. 1995. Physical Health Symptomatology of Gulf War-Era Service Personnel from the States of Pennsylvania and Hawaii. *Military Medicine* 160:131–136.

Stretch, R.H., Bliese, P.D., Marlowe, D.H., Wright, K.M., Knudson, K.H., and Hoover, C.H. 1996a. Psychological Health of Gulf War-Era Military Personnel. *Military Medicine* 161:257–261.

Stretch, R.H., Marlowe, D.H., Wright, K.M., Bliese, P.D., Knudson, K.H., and Hoover, C.H. 1996b. Post-Traumatic Stress Disorder Symptoms Among Gulf War Veterans. *Military Medicine* 161:407–410.

Sudman, S., Bradburn, N., and Schwarz, N. 1996. *Thinking About Answers: The Application of Cognitive Processes to Survey Methodology*. San Francisco, Calif.: Jossey-Bass.

Sugarbaker, P.H., Barofsky, I., Rosenberg, S.A., and Gianola, F.J. 1982. Quality of Life Assessment of Patients in Extremity Sarcoma Clinical Trials. *Surgery* 91(1):17–23.

Sugman, S., and Bradburn, N. 1982. *Asking Questions: A Practical Guide to Questionnaire Design*. San Francisco, Calif.: Jossey-Bass.

Sutker, P.B., Davis, J.M., Uddo, M., and Ditta, S.R. 1995. Assessment of Psychological Distress in Persian Gulf Troops: Ethnicity and Gender Comparisons. *Journal of Personality Assessment*. 64:415–427.

Sutker, P.B., Uddo, M., Brailey, K., Vasterling, J.J., and Errera, P. 1994. Psychopathology in War-Zone Deployed and Nondeployed Operation Desert Storm Troops Assigned Graves Registration Duties. *Journal of Abnormal Psychology* 103:383–390.

Thomas, M.R., and Lyttle, D. 1980. Patient Expectations About Success of Treatment and Reported Relief from Low Back Pain. *Journal of Psychosomatic Research* 24(6):297–301.

Torrance G.W., and Feeny D. 1989. Utilities and Quality-Adjusted Life Years. *International Journal of Technology of Assessment in Health Care* 5:559–578.

U.S. Senate Committee on Banking, Housing, and Urban Affairs. 1994. *United States Dual-Use Exports to Iraq and Their Impact on the Health of the Persian Gulf Veterans*. Senate Report 103-900, May 25, 1994.

U.S. Senate Committee on Veterans' Affairs. 1998. *Report of the Special Investigation Unit on Gulf War Illnesses*. Washington, D.C.: U.S. Government Printing Office.

Unwin, C., Blatchley, N., Coker, W., Ferry, S., Hotopf, M., Hull, L., et al. 1999. Health of U.K. Servicemen Who Served in Persian Gulf War. *Lancet* 353:169–178.

Ware, J.E., Snow, K.K., Kosinski, M., and Gandek, B. 1993. *SF-36 Health Survey: Manual and Interpretation Guide*. Boston: Nimrod Press.

World Health Organization (WHO). 1948. *Constitution of World Health Organization. Basic Documents*. Geneva: WHO.

Wolfe, J., Proctor, S.P., Davis, J.D, Borgos, M.S., and Friedman, M.J. 1998. Health Symptoms Reported by Persian Gulf War Veterans Two Years After Return. *American Journal of Industrial Medicine* 33:104–113.

Wood, R.W., Diehr, P. Wolcott, B.W., Slay, L., and Tompkins, B.K. 1979. Reproducibility of Clinical Data and Decisions in the Management of Upper Respiratory Illnesses. *Medical Care* 17:767–779.

Writer, J.V., DeFraites, R.F., and Brundage, J.F. 1996. Comparative Mortality Among U.S. Military Personnel in the Persian Gulf Region and Worldwide During Operations Desert Shield and Desert Storm. *Journal of the American Medical Association* 275:118–121.

APPENDIX A

Coalition Forces and Force Strength

Total U.S. forces deployed: 697,000; peak personnel strength: 541,400
Total other coalition forces: 259,700 at peak personnel strength

Country	No.	Country	No.
Afghanistan	300	Norway	50
Argentina	300	Oman	6,300
Australia	700	Pakistan	4,900
Bahrain	400	Philippines	Unknown
Bangladesh	2,200	Poland	200
Belgium	400	Qatar	2,600
Canada	2,000	Romania	Unknown
Czechoslovakia	200	Saudi Arabia	100,000
Denmark	100	Senegal	500
Egypt	33,600	Sierra Leone	Unknown
France	14,600	Singapore	Unknown
Germany	Unknown	South Korea	200
Greece	200	Spain	500
Hungary	50	Sweden	Unknown
Italy	1,200	Syria	14,500
Japan	Unknown	Thailand	Unknown
Kuwait	9,900	Turkey	Unknown
Morocco	13,000	United Arab Emirates	4,300
Netherlands	600	United Kingdom	45,400
New Zealand	Unknown	Zaire	Unknown
Niger	600		

APPENDIX B

Design Issues in the Gulf War Veterans Health Study

Naihua Duan
RAND, Santa Monica, California
and
Robert O. Valdez
School of Public Health, University of California at Los Angeles

ABSTRACT

The design for the Gulf War Veterans Health Study poses a variety of challenges. In order to study changes in the Gulf War (GW) veterans' health status over time, a panel design (also known as the prospective cohort design) is indicated. Since the study also aims to examine the levels of the GW veterans' health status at various time points, consideration needs to be given to the cross-sectional representativeness of the panel, thus a repeated panel design should be considered as a potential alternative to a permanent panel design. Due to the anticipated deterioration in the quality of the locating information on the GW veterans, recruiting the GW veterans will likely require a substantial effort to track and trace the sampled participants, making it unattractive to use designs such as a rotating panel that requires repeated recruitment of new panels. Given the closed nature of the GW veteran population (there are no new entries), it is important for the study to provide timely information of value to the GW veterans during their lifetime. Thus, the study should be designed with more frequent data collection in the early years when the information obtained has a longer "useful life." Based on the consideration of various trade-offs, three of the most promising designs are the permanent panel design, the repeated panel design, and a combination of the two. A promising design is to recruit an initial panel and follow the panel every 3 years for three waves. An assessment shall be made after the third wave to assess the quality of the panel, to determine whether to continue following the same panel, to switch to a new panel, or to take a combination of the two. The survey frequency might be reduced in the second decade and beyond.

Prepared for the Institute of Medicine Committee on Measuring the Health of Gulf War Veterans.

INTRODUCTION

The design of a study is usually determined by the research questions that need to be addressed, and the target population to be studied. The proposed Gulf War Veterans Health Study (GWVHS) aims to address the following research questions:

1. How healthy are Gulf War veterans?
2. In what ways does the health of Gulf War veterans change over time?
3. Now and in the future, how does the health of Gulf War veterans compare with:

- The general population?
- Persons in the military at the time of the Gulf War but not deployed?
- Persons in the military at the time of the Gulf War who were deployed to nonconflict areas?
- Persons in the military deployed to other conflicts, such as Bosnia, Somalia, and so on?

4. What individual and environmental characteristics are associated with observed differences in health between Gulf War veterans and comparison groups?

Since the study aims to address both the levels at specific time points (the first and third research questions) and changes over time (the second and fourth research questions) for Persian Gulf veterans' health status, we will consider study designs appropriate for both types of research questions. In particular, we will consider repeated cross-sectional surveys and various panel survey designs (also known as prospective cohort designs).

Unlike the general population (either civilian or military), Gulf War (GW) veterans are a closed population: there is no birth, enlistment, or migration into this population. The membership in this population was determined by the participation in the Persian Gulf War, that is, those who served in the Gulf War theater of operations between August 2, 1990, and June 13, 1991. The closed nature of the GW veteran population has important implications for the study design, such as on the merits of replenishing the panel.[1] Further discussions are given in the section on temporal structure below.

Given the closed nature of the GW veteran population, it is important for the GWVHS to provide timely information for the GW veterans. This objective gives the GWVHS a stronger focus on public health rather than basic science research. The information obtained in GWVHS will be of value to the GW veterans only during their lifetime. Therefore timeliness of information should be

---

[1]Discussions about the use of refreshment samples in a panel study with attrition are given in Hirano et al. (1998).

taken into consideration for the design of the GWVHS. In other words, the focus for GWVHS is more on public health (to serve the GW veterans) than basic science research (to obtain scientific knowledge applicable to future patients). This important feature indicates that the GWVHS should be designed with more frequent data collection in the early years when the information obtained has a longer "useful life" to the GW veterans, and less frequent data collection in later years when the information has a shorter "useful life." Further discussions are given in the section on Survey Frequency.

In order to strengthen our ability to understand the health problems of GW veterans and their trajectories, it is worth considering a case-control component in GWVHS. The comparison group would consist of patients with medically unexplained physical symptoms in the same geographical areas as the GW veterans in the GWVHS sample. Geographical matching has an important implication for the study design, namely, the extent to which the GWVHS sample should be clustered geographically. Further discussions are given in the section on Survey Modality and Geographical Clustering.

Another important unique feature about the GW veteran population is that our ability to locate individuals in this population is likely to deteriorate over time. While the Department of Defense (DoD) maintains locating information on record for all veterans, including GW veterans, the accuracy of this information is likely to decrease over time. As this report is being written, 8 years have elapsed since the Gulf War. We anticipate, therefore, that a substantial tracking and tracing effort is necessary to recruit a representative sample of GW veterans, and a substantial level of nonresponse will still occur despite this effort. This feature has important implications for the design of GWVHS, making it less appealing to recruit multiple cohorts into the study.

In population studies of Gulf War veterans conducted to date, response rates ranged from a low of 31% in the study conducted by Stretch et al. (1995) to 97% of those located in a survey of women who served in the U.S. Air Force during the Gulf War conducted by Pierce (1997). Further details are given in Chapter 5 of the report. A low response rate is a concern for the validity of the data, therefore it is important to engage in efforts to increase the response rate, using survey research tools such as tracking and tracing of participants, and incentives. Further discussions on nonresponse, tracking, and tracing can be found in the section "Nonresponse, Attrition, Tracking, and Tracing."

In order to help interpret the health status of GW veterans (especially the changes over time), several comparison groups (listed under the third research question above) will be included in GWVHS. Those comparison groups will be recruited and surveyed using the same design to be used for GW veterans to maximize the comparability. Since the membership in the GW veteran population versus the comparison groups is not randomized, the comparison will be vulnerable to potential selection bias problems common to all observational studies: GW veterans might be different from the comparison groups even in the absence of the Gulf War experience. In order to account for such differences, the best possible effort needs to be made to collect data on potential confounding factors.

Based on the consideration of various trade-offs in the ability of various study designs to accomplish the objectives for the GWVHS, three of the most promising designs are the permanent panel design, the repeated panel design, and a combination of the two. A promising design is to recruit an initial panel and follow them every 3 years for three waves. An assessment shall be made to evaluate the quality of the panel at the end of the third wave, to determine whether to continue following the same panel, to switch to a new panel starting in year 10,[2] or to take a combination between the two. The survey frequency might be reduced in the second decade and beyond. Note that the final design decision can be made after the first three waves have been fielded; thus it can be (and should be) based on the actual cost data obtained in the field.

In order to facilitate the recruitment of the second panel if warranted, it is worth considering recruiting a "reserve" sample along with the initial panel, giving them a brief enrollment interview to collect contact information, and maintaining contact with them over time through tracking. This provision will reduce the potential deterioration in our ability to locate and contact the GW veterans who were not sampled in the initial panel.

TEMPORAL STRUCTURE FOR SURVEY STUDIES

Given the dual goals in GWVHS to study both levels and changes for GW veterans' health status, it is important to consider the structure of the survey study design both in terms of the units (individuals) to be surveyed and the time(s) those units are to be surveyed. We discuss in this section the candidate designs, the pros and cons for those designs, and specific considerations for the GWVHS.

Taxonomy of Survey Studies

The temporal structure for survey study design can be classified according to the following taxonomy,[3] listed in order of increasing emphasis on temporal versus unit-specific data collection.

---

[2]Starting the second panel a year after the end of the first panel, instead of following the regular 3-year interval between waves, would facilitate the opportunity to "splice" across the panels.

[3]Similar taxonomies and discussions about the trade-offs among those designs are given in Duncan and Kalton (1987), Bailar (1989), and Kalton and Citro (1993). General discussions about design and analysis issues for panel studies are also given in Duncan et al. (1984) and Hsiao (1985, 1986).

1. *Single cross-sectional survey.* A cross-sectional sample of observation units is identified at one time point and surveyed once. This design only allows for the estimation of level parameters, such as the prevalence of medically unexplained physical symptoms, at the specific time point, and usually does not allow for the assessment of prospective changes over time. The survey might include retrospective data items to inquire about past events; thus, it might provide some information on past changes, although the reliability of the retrospective data might be compromised. Additionally, the validity of retrospective data might be affected by mortality and other forms of exit from the target population. For example, in order for patients who experienced a life-threatening disease to be available to report on past changes, they must have survived the disease. Therefore the survival rate estimated from the retrospective data is likely to be very biased.

2. *Repeated cross-sectional surveys without unit overlap.* A cross-sectional sample of observation units is identified at each of several time points; each sample is surveyed once. The samples are drawn with no provisions for overlap; a few overlap cases might occur if the samples are drawn with replacement and the sampling rates are sufficiently high. This design allows for the estimation of level parameters at each of the selected time points, as well as the average of the level parameters over time. In addition, this design also allows for the estimation of net changes in the population parameters over time, such as the increase or reduction of the prevalence of medically unexplained physical symptoms over time. It usually does not allow for the assessment of individual gross changes, such as the persistence of these unexplained symptoms, whether the same patients are affected over time, and the amount of turnover. (Retrospective inquiries might provide some information, but might be of compromised quality.)

3. *Repeated cross-sectional surveys with unit overlap.* This design is similar to (2), with the exception that a portion of each subsequent sample is drawn from the previous sample. In other words, the membership in the previous sample is used as a stratifying variable in the subsequent sample, with the units in the previous sample being oversampled relative to the new units. This design has similar capabilities as (2); it also allows for the estimation of gross changes on the individual level, using the portion of the sample that overlaps with the previous sample.

4. *Repeated panel surveys with temporal overlap.* This design is similar to (2), with the exception that each sample is surveyed several times, usually at regular time intervals, thus each sample serves as a panel. The temporal spans for the panels overlap in time: a new panel is initiated before the previous panel has retired, thus several panels are usually active in the field at the same time. This design is essentially the same as the rotating panel survey design. (There are some fine distinctions between those two designs, but the similarity dominates their differences.) For each time point, the level parameters can be estimated using the panels active at the time, usually including a new panel (just initiated) and several ongoing panels. The quality of the ongoing panels for the level estimates might be compromised by panel artifacts such as attrition and

panel conditioning, to be discussed later. (Similar compromise might also occur in [3] among the overlap portion of the subsequent samples.) However, the presence of multiple panels in different stages of progression might help mitigate some of those problems. On the other hand, this design usually provides more precise estimates on the changes (both net and gross) than the previous designs.

5. *Repeated panel surveys without temporal overlap*. This design is similar to (4), with the exception that the panels do not overlap in time: a new panel is initiated after the previous panel is retired. (Thus the temporal span for each panel is distinct and does not overlap with other panels.) Each panel allows the estimation of level parameters at the time of its initiation, as well as the estimation of those parameters at each follow-up. Similar to (4), this design might also be vulnerable to attrition and panel conditioning, and is likely more vulnerable than (4) at follow-up waves because there are no other panels active at the same time to help mitigate those problems.

6. *Permanent panel survey*. A single sample is drawn at one specific time point, then followed for the entire duration of the study. This design provides more information on changes, especially on long-term gross changes on the individual level. (The ability of [4] and [5] to provide direct information on long-term gross changes is limited by the duration of the individual panels. It might be possible to "splice" distinct panels to assess long-term gross changes. This usually requires strong assumptions such as the Markovian properties on the nature of the gross changes.) On the other hand, this design is more vulnerable than (4) and (5) to attrition and panel conditioning.

7. *Time series study*. A single unit is chosen and followed intensively for the entire duration of the study. This design will provide the most intensive information on gross changes, and practically no information on level parameters.

There are many possible variations on and combinations of the designs listed above. For example, a study focused on a specific disease or condition might follow all cases with the condition to assess their trajectories, and follow a subsample only for the noncases to assess the incidence rates.

Pros and Cons for Alternative Study Designs

Given the dual goals for GWVHS to obtain both level and change estimates, designs (1) and (7) should be ruled out. Among the remaining candidate designs (2)–(6), the sequence in which they appear in the above section on Taxonomy is ranked in increasing order of the emphasis on repeated measurements on the same individuals, that is, the overlap of the cohort over time.

Generally speaking, the more overlap there is across time in the units surveyed, the more information is available on estimating changes. This premise is self-evident for gross changes on the individual level: we observe gross changes only among the units that overlap in time.[4] The permanent panel survey is therefore the preferred design for studies focused on gross changes.

For net changes, the overlap is usually viewed as a plus because the same individuals serve as their own controls over time. This is illustrated in the following simple model:

$$Y_{it} = \alpha + t\beta + \theta_i + \varepsilon_{it}, \quad i = 1,\ldots,n; \quad t = 0,1,$$

where Y denotes the outcome of interest, α denotes the baseline population status, β denotes the net change of interest, θ denotes the time-invariant individual variation, and ε denotes the temporal random error. With less overlap, such as in repeated cross-sectional surveys, distinct individuals are observed at times 0 and 1. The net change is estimated using the difference between the two sample means:

$$\hat{\Delta}_{CS} = \overline{Y}_1 - \overline{Y}_0 = \beta + (\overline{\theta}_1 - \overline{\theta}_0) + (\overline{\varepsilon}_1 - \overline{\varepsilon}_0),$$

where the subscripts 0 and 1 refer to the distinct cross-sectional samples. The uncertainty in the estimated net change includes both the sampling error in the cross-sectional samples, θ, and the temporal random error, ε. With more overlap, such as in panel surveys, the same individuals are observed at times 0 and 1, thus the sampling error θ is cancelled when we compare times 0 and 1, resulting in a more precise estimate for the net change.[5]

For estimating levels, the overlap is usually a disadvantage.[6] More specifically, consider the estimation of the average level of Y across the population and also over time. This is usually estimated using the grand mean of all observed Y_{ij}'s. With repeated cross-sectional surveys, the sampling error variance is reduced by a factor of $2n$, because two distinct (and independent) samples are drawn at times 0 and 1. For panel survey, the sampling error variance is reduced by a factor of n, because the same sample is used at times 0 and 1. Therefore the

[4] It is possible to use repeated cross-sectional surveys to estimate gross change parameters such as survival rates under various assumptions. The precision for those estimates is usually substantially lower compared to those obtained using panel surveys.

[5] For simplicity of illustration, we restricted ourselves here to a two-wave design and an analysis of changes. With more waves of data collected on the same participants, a rich variety of longitudinal analysis (time trend analysis, growth curve analysis) can be applied (see, e.g., Hsiao, 1985, 1986; Diggle et al., 1994). Discussions on the pros and cons of cohort designs versus repeated cross-sectional designs for community intervention studies are also given in Diehr et al. (1995) and Gail et al. (1996).

[6] An important implication is that the less overlapped designs such as the repeated cross-sectional surveys without unit overlap are more capable of identifying cases with rare attributes such as rare disease conditions.

estimate based on the panel survey is less precise.[7] This comparison is important, for example, for detecting persistent rare conditions: the chance for detecting such conditions is much higher with repeated cross-sectional surveys than panel surveys (because more individuals are surveyed).

A related advantage for the panel survey design is its ability to control for time-invariant *unobserved* confounding factors. As an illustration, consider the following extension of the earlier model:

$$Y_{it} = \alpha + t\beta + x_{it}\gamma + w_i\delta + \theta_i + \varepsilon_{it}, \quad i = 1,\dots,n; \quad t = 0,1,$$

where x denotes the predictors of interest, and w denotes unobserved confounding factors (assumed to be time-invariant). If the confounding factors w were observed, it would be possible to control for them in cross-sectional data. With panel data, it is possible to control for unobserved time-invariant confounding factors by taking the difference across time, resulting in the following difference model:

$$\Delta_i = Y_{i1} - Y_{i0} = \beta + (x_{i1} - x_{i0})\gamma + (\varepsilon_{i1} - \varepsilon_{i0}), \quad i = 1,\dots,n.$$

We then regress the change in Y (Δ) on the change in x. Assuming that the confounding factors w are time-invariant, they would be cancelled out in the difference model.

It should be noted, though, that the ability of the difference model to control for confounding factors and estimate the effects of interest depends critically on the temporal variation in the predictors of interest. If the predictors x do not vary over time, the difference model does not allow us to estimate their effects. Even if the predictors x do vary over time, the precision for the estimated effects might be poor if the temporal variation in x is small.

Another important advantage of the panel design is that it might help improve the quality of the recall by using the events observed in the earlier waves to bound the time frame of recall in future waves (see, e.g., Neter and Waksberg, 1964). However, since the anticipated between-wave lags are much longer than the recall period for GWVHS, this feature is unlikely to be applicable. Further discussions on bounding the time frame and related recall error issues can be found in the section, Measurement Error.

While the panel design has much merit, it also has some limitations. One important limitation is that the panel design can be especially vulnerable to nonresponse. Nonresponse is an important limitation to all survey studies under both cross-sectional and panel designs. Almost all survey studies fail to obtain complete data on some sampled subjects due to various reasons: some subjects cannot be located or reached, some are too sick to be interviewed, some refuse to be

[7] This comparison assumes the same sample size under the two designs. This is somewhat unfair: the sample size available under the panel design will likely be larger due to its lower cost per person-wave.

interviewed. It is usually plausible that the nonrespondents are different from the respondents in terms of the attributes of interest. For example, Groves and Couper (1998) reported that households with many members and households with elderly persons or young children are easier to contact, urban households are more difficult to contact than rural households; once contacted, those in military services, racial and ethnic minorities, and households with young children or young adults are more likely to cooperate with the surveys. Given the potential for respondents to differ from nonrespondents, the analyses based on the respondents might provide biased estimates for the target population. The severity of the nonresponse bias is usually associated with the nonresponse rate (see, e.g., Kish, 1965, Section 13.4B). If the nonresponse rate is low, say, less than 10%, the nonresponse bias is likely to be small and negligible. If the nonresponse rate is high, say, more than 30%, there is a potential that the nonresponse bias might be serious, thus the conclusions based on the respondents might be flawed.

There are a number of statistical and econometric techniques that can be used to mitigate the impact of nonresponse, such as nonresponse weighting, (multiple) imputation, pattern mixture modeling, and selectivity modeling. The section Nonresponse, Attrition, Tracking, and Tracing provides further discussion.

While nonresponse is usually a limitation for both cross-sectional and panel designs, it is usually a bigger problem for panel designs because the nonresponse can accumulate over time. A panel study that is designed and implemented well usually holds the attrition over time to a very low level, such as 5 to 10 percent in each wave. Furthermore, some sampled subjects who did not respond to an early wave might be "resurrected" in a later wave. However, the nonresponse usually accumulates across waves, and reaches a substantial level after multiple waves. For example, the wave nonresponse accumulated to 19% at wave 7 in the 1987 Survey of Income and Program Participation (SIPP) panel (Jabine et. al., 1990), one of the exemplary panel studies. Therefore, the potential for nonresponse bias becomes more severe in later waves.[8]

A related limitation for the panel design is the omission of new members in the target population, due to birth, enrollment, immigration, and so on. While the panel was designed to be representative of the target population at the beginning of the study, the panel ages over time and does not represent the new members. Therefore the representativeness of the panel becomes compromised in subsequent waves, both because of the omission of new members, and because of the cumulative attrition discussed earlier. Some panel studies refresh the sample by adding a sample of new members who joined the target population since the original sample was drawn. This can be costly unless there is an easy way to identify the new members. However, since the GW veterans are a closed population, there are no new members entering the target population, thus the omission of new members is not an issue.

---

[8]The ability of the statistical techniques to mitigate the potential nonresponse bias also improves over time with the panel design, because we have more data on the sampled subjects who responded to earlier waves and drop out in a later wave.

Another important limitation in the panel design is panel conditioning: the observed responses might be affected by the participation in the panel, thus compromising the validity of the data obtained in later waves. Further discussions on panel conditioning can be found in the section, Measurement Errors Specific to Panel Surveys.

GULF WAR VETERANS HEALTH STUDY DESIGN

Repeated panel surveys with temporal overlap (rotating panel surveys) is the commonly used compromise when both levels and changes are of interest. However, this design requires recruiting new cohorts from the target population regularly; thus, it might not be appropriate for the GWVHS. Given the anticipated deterioration of the quality of the DoD records data on the GW veterans, the recruitment cost is likely high for the GW veteran population, requiring substantial tracking and tracing efforts. In order to economize the design, it would be desirable to reduce the need to recruit new cohorts. Based on those considerations, either a permanent panel design or repeated panel surveys without temporal overlap would be the preferred choice for GWVHS, to avoid conducting costly recruitment on a regular basis.

The permanent panel design has the advantage that it allows the direct assessment of long-term gross changes on the same individuals—with the repeated panel surveys without temporal overlap, we need to "splice" the trajectory from different panels to assess changes across waves that fall under different panels. However, the permanent panel design will be more vulnerable to cumulative attrition and panel conditioning. Therefore, a promising design also worth considering for the GWVHS is repeated panel surveys without temporal overlap. The study should review the quality of the first panel after the third wave for the initial panel,[9] to determine the extent to which the validity of the inference based on the panel is compromised by cumulative attrition and panel conditioning.[10] If the quality of the panel is judged to be satisfactory, the study would continue following the same panel. If the validity is judged to be unsatisfactory, the study would switch to a new panel. If the validity is judged to be marginal, it is con-

[9]Cumulative attrition usually levels off after the first two or three waves in the existing panel studies. Therefore it is reasonable to conduct the assessment after the third wave, and assume that further attrition will be small. It is reasonable to assume that panel conditioning will also level off after two or three waves, although there is less empirical evidence.

[10]The assessment of cumulative attrition is straightforward. The assessment of panel conditioning is more involved, and requires differentiating the true changes over time from panel conditioning. Ideally, the assessment should use a new sample, and compare the distribution of outcome measures between the new sample and the on-going panel. However, since we anticipate that panel conditioning is unlikely to have a major impact on the GWVHS participants, it might not be worthwhile to devote a substantial amount of resources to conduct this assessment.

ceivable that a hybrid design analogous to a rotating panel design could be used, continuing to follow a random subsample of the initial panel, and drawing a new panel to make up for the discontinued portion of the initial panel.

In order to facilitate the recruitment of a second panel if it is warranted, it is worth considering that a "reserve" sample be recruited at the same time of the initial sample. The "reserve" sample will be enrolled into the study, and given a brief survey to collect the contact information.[11] This sample will then be sent into "hibernation," and will be reactivated if a decision is reached later to recruit a second panel. While the "reserve" sample is in "hibernation," we will maintain tracking to make it feasible to reactivate this sample if needed.[12] This provision will require a nontrivial amount of resources, but will guard against the risk that the contact information in the DoD records will deteriorate further during the tenure of the initial panel, making it impossible to recruit a second panel.

If the "reserve" sample is implemented successfully, the rotating panel design might be a viable option. We can activate a third of the "reserve" in waves 2, 3, and 4, respectively, and retire a corresponding portion of the original panel. The maintenance cost for the "reserve" sample will be lower under the rotating panel design, because the size of the "reserve" sample is reduced over time. On the other hand, the recruitment cost is still likely to be substantial even with the "reserve" sample, therefore it might be more economical to activate the "reserve" sample in one lump sum instead of in pieces.

There are important analytic trade-offs between those designs. With the repeated panel design without temporal overlap, the entire panel is available in the first three waves, allowing more precise estimates for changes (both net and gross). Furthermore, this design allows the option of switching to a permanent panel design (without activating the "reserve" sample), either in part or in full, if warranted. However, we cannot estimate gross changes between the two panels, say, between the third and fourth waves. On the other hand, the rotating panel design includes less overlap across the first three waves; thus, it provides less precision for estimating changes (especially gross changes) among those waves. However, it does allow for the estimation of gross changes in later waves, say, between the third and fourth waves. As discussed in the Introduction to this appendix, the public health focus of the GWVHS indicates that it is more important to assess changes across the first three waves (repeated panel surveys without temporal overlap is preferable for those objectives), than assessing changes that occur in later waves (rotating panel design is preferred for those objectives). See Survey Frequency for further discussion.

---

[11]We might as well collect a minimal amount of health status data at the same time the contact information is collected.

[12]Since the "reserve" sample will be in "hibernation" for many years before reactivation, the tracking should be conducted in a cost-effective way, using the low-cost procedures only. A more comprehensive tracing effort will be conducted when the sample is reactivated.

Another analytic advantage in the rotating panel design is that it allows us to assess the presence and severity of panel conditioning regularly because a new portion of the "reserve" sample is activated in each wave. However, given our anticipation that panel conditioning is unlikely to be a major problem for the GWVHS (see Survey Modality and Geographic Clustering), this advantage is less critical. In summary, the challenges in recruiting the GW veterans make the rotating panel design less appealing for the GWVHS than either the permanent panel design or the repeated panel design without temporal overlaps.

To summarize, three of the most promising designs for the GWVHS are the permanent panel design, the repeated panel design, and a combination of the two. A promising design is to recruit an initial panel, and follow them every 3 years for three waves. An assessment shall be made to evaluate the quality of the panel at the end of the third wave (to be fielded during the 9th year of the study), to determine whether to continue following the same panel, to switch to a new panel starting in year 10, or to take a combination between the two. It is worth considering recruiting a reserve sample to be activated as needed for the second panel. Note that the final design decision can be made after the first three waves have been fielded; thus, it can be (and should be) based on the actual cost data obtained in the field.

SURVEY FREQUENCY

An important parameter in the design of panel studies is the survey frequency: how often will the participants be surveyed. We assume for now that the survey is conducted with equal spacing between waves—this is the design commonly employed in most panel studies. We also assume that the overall duration of the study has been determined in advance. The survey frequency is then determined by the number of waves to be conducted within the given duration.

For many panel studies, the survey frequency is determined by the reference periods for the key outcome measures (see the section, Measurement Error below), so as to obtain a contiguous stream of those outcome measures over time. Since the GWVHS follow-up waves will be much longer than the reference periods for the outcome measures, this aspect is not applicable.

For studies with a fixed budget, there is usually a trade-off between the survey frequency and the number of participants surveyed in each wave. With a higher survey frequency, the number of participants surveyed in each wave will be smaller.

For panel studies, a simplistic way to measure the total amount of information collected can be measured using the number of person-waves of surveys conducted. The cost per person-wave surveyed usually decreases with the survey frequency. More specifically, the marginal cost to conduct an additional wave of the follow-up survey for a participant already recruited into the study is usually lower than the cost to recruit an additional participant into the study for the first wave of data collection, because tracking and tracing costs for ongoing participants are usually lower than the recruitment cost for new participants. Therefore, this simplistic measure (the number of person-waves surveyed) usually favors collecting many waves of follow-up on the same sample of participants.

The simplistic measure described above does not convey all information relevant to the design of a panel study. We must also consider the statistical information obtained in the study. Due to intraclass correlation at the individual level, the statistical information per person-wave of survey is likely to decrease with the survey frequency, thus counter-balancing the reduction in the cost per person-wave of survey. The intraindividual correlation can occur on the levels: a healthy individual will usually remain healthy over time. The correlation can also occur on the changes: the trajectory of an individual's health change during, say, the first 3 years might be predictive of his/her trajectory in the future.

As was noted in the Introduction, the repeated measurements on the same participants usually contribute less information towards the estimation of cumulative levels, such as the detection of rare conditions. For estimating changes, the panel design is usually preferable to the repeated cross-sectional design. However, the statistical information does not accumulate proportionately with the number of waves surveyed. Assuming a linear time trend, the largest contribution to the statistical information of the estimated rate of change comes from the first and last waves. The marginal contribution from each intermediate wave is smaller.

The appropriate survey frequency needs to balance between the cost per person-wave of survey and the statistical information per person-wave of survey. Overall and Doyle (1994) and Hedeker, Gibbons, and Waternaux (1999) provided the methodologies for power calculations for longitudinal studies under a variety of model assumptions. Those techniques can be used in conjunction with information about survey costs to compare alternative survey frequencies to determine the appropriate design that provides the optimal power and precision under the available budget.

In addition, one also needs to take into consideration attrition and measurement error issues discussed in the section Nonresponse, Attrition, Tracking, and Tracing and in Measurement Error, in order to arrive at the appropriate choice for the survey frequency. This is not an easy task. Presser (1989) observed that "There are difficult trade-offs here . . . a given decision might decrease sampling error but increase potential nonresponse bias, while at the same time reducing conditioning and increasing telescoping . . . the paucity of relevant studies means we are frequently operating in the dark."

A useful recommendation was given in Cantor (1989) to base decisions on survey frequency on "the amount of time it takes to expect meaningful change and/or occurrence in the variables that are of substantive interest. . . . For example, epidemiological studies typically require a number of years for a follow-up period in order to allow for physical and mental development." In other words, the survey frequency should allow enough time between waves for meaningful changes worth measuring to take place.

We have assumed up until now that the survey is conducted with equal spacing between waves—this is the design commonly employed in most panel studies. This might not be the most appropriate design for the GWVHS, for several reasons.

First, the usual practice of conducting longitudinal studies with equal spacing is not well-grounded in statistical theory. Maxwell (1998) observed that "equal spacing of observation periods is a typical mathematical assumption, in part because many longitudinal designs follow this practice. Nevertheless, if the straight-line grow model is in fact the correct model, greater power can be achieved by spacing all of the assessments farther from the middle of the overall assessment period and closer to the extremes (i.e., closer to the pretest and to the posttest). On balance, the equally spaced design often offers a better choice in practice. . . ."

The observation that nonequal spacing can improve power is important. In the extreme, if there is compelling evidence that the change in the outcomes follows the straight-line model (the rate of change is constant over time), the intermediate waves do not contribute much information, thus the study design can be improved by placing more observations near the beginning and the end of the study period. The optimality under the straight-line model needs to be balanced against the robustness for the design to perform well under deviations from the straight-line mode, thus some waves should be placed in the intermediate time range. However, as long as the straight-line model is a reasonable approximation to the actual pattern of the change over time, the balance between optimality and robustness is unlikely to justify the equally spaced design—the appropriate design should still place more emphasis near the beginning and the end of the study period.

In addition, the public health (vs. basic science research) focus of the GWVHS makes it important to take into consideration the timeliness of interim data. As discussed in Section 1, the information obtained in GWVHS will be of value to the GW veterans only during their lifetime. Therefore timeliness of information should be taken into consideration for the design of the GWVHS.

In basic science research, we usually assume an infinite time horizon.[13] The scientific knowledge acquired from the study is expected to serve a large number of future clients in the years to come. The study participants might receive suboptimal treatments in order to provide the scientific knowledge to serve future clients. For example, in placebo-controlled trials, the participants assigned to the control group receive the placebo instead of the experimental treatment.

---

[13]Weinstein (1974) criticized the assumption of an infinite time horizon in clinical trials, and advocated considering a limited time horizon instead. Indeed, the "useful life" of scientific knowledge is usually limited; a new treatment will likely be replaced by other treatments in several years. A variety of alternatives to the usual clinical trial design have been proposed in the literature, such as sequential designs, two-stage designs, and play-the-winner designs, but those designs are not widely used (see, e.g., Armitage, 1960; Coad and Rosenberger, 1999; Day, 1969; Lai et al., 1980; Robbins, 1974; Wei and Durham, 1978; Whitehead, 1997; and Zelen, 1969).

Although the ethical consideration dictates that the trial be conducted only in the absence of *conclusive* evidence favoring the experimental condition over the control condition, it is conceivable that some partial evidence is available to justify the plausible benefits in the experimental condition. During the course of the trial, interim comparisons between the conditions are usually made to monitor the accumulation of evidence. The trial will usually be terminated when *conclusive* evidence is found favoring either the experimental or the control condition. However, it is conceivable that the interim data might indicate weak but *not-yet-conclusive* evidence; the trial usually continues, to allow further accumulation of evidence. In essence, the trial participants are making a short-term sacrifice to contribute to the benefits of future patients. The performance of the trial is usually judged based upon the value of the scientific knowledge acquired at the end of the study. Interim evidence is usually not used for the benefits of the trial participants except in extreme cases when the interim evidence is *conclusive*. Timeliness of information (the value of the interim data) is usually not taken into consideration for study designs because the dissemination phase subsequent to the trial has an infinite time horizon.

On the other hand, a public-health-oriented study like GWVHS aims to serve today's target population; thus, it does have an infinite time horizon.[14]

The following diagram illustrates the pattern of information accumulation in a basic science research study that attempts to estimate the rate of change over time (assumed to be constant) in the outcome variable, using five waves of data collection.

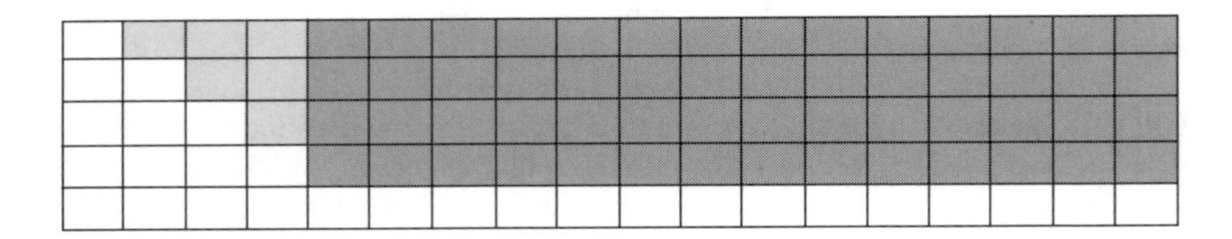

The horizontal axis in the diagram denotes time. The baseline survey is conducted at time zero, corresponding to the left edge of the diagram. The first follow-up is conducted at time one, corresponding to the first vertical line inside the diagram. The second follow-up is denoted by the next vertical line, and so on. As an illustration of the basic science research paradigm, we assume that the study period is relatively short compared to the dissemination period during which the information obtained in the study is utilized.

The vertical axis denotes the level of uncertainty about the rate of change. Prior to the first follow-up, the study provides no information about the rate of change. The first white bar between time zero and time one denotes the level of

[14]It is conceivable that the GWVHS can also serve a basic science function and provide scientific knowledge applicable to veterans in future wars and conflicts. We assume that this objective is secondary to the objective of serving the GW veterans.

uncertainty based on prior information. The first follow-up provides some new information about the rate of change, thus reducing the uncertainty somewhat. The second white bar between times one and two denotes the level of uncertainty remaining after the first follow-up. The portion of the second bar shaded in light gray denotes the information gain (the reduction in uncertainty) due to the first follow-up. Further reduction in uncertainty is accomplished with each follow-up, until the fourth and final follow-up. The white space subsequent to time four denotes the uncertainty remaining after the conclusion of the study. (The reduction in uncertainty is not linear in time; the diagram should be viewed as a conceptual demonstration.)

The gray areas in the diagram denote the reduction in uncertainty. The portion of the reduction due to the interim data is shaded in light gray. The portion of the reduction available after the conclusion of the study is shaded in dark gray. Since the study period is short compared to the dissemination period, the value of the interim data is low compared to the value of the final study results available after the conclusion of the study.

The next diagram illustrates a public-health-oriented study like the GWVHS. This study has the same five waves of data collection. However, the time horizon is much shorter than the previous study. The objective of this type of study is to serve the current population being studied, rather than to serve future populations. As can be seen from the diagram, the value of the interim data is high compared to the value of the final study results. Indeed, if the GWVHS follows the GW veterans until the end of their life cycles, the final study results will be of very little value to the GW veterans.

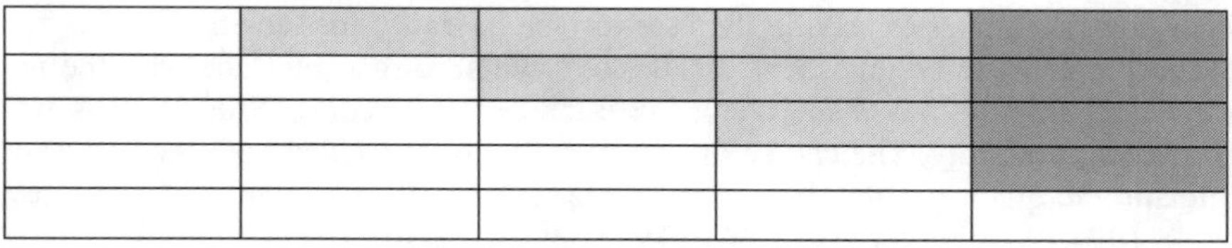

Under the paradigm discussed above, there is an important rationale to place more emphasis on data collection in the early years in the GWVHS: this information can be utilized to serve the GW veteran population over a longer time period that the data collected in later years. It appears reasonable that a higher survey frequency should be used in the early years in the GWVHS to maximize the overall utility of the information obtained over time, as illustrated in the diagram below. Note that we have varied both the spacing between waves, and the amount of uncertainty reduction corresponding to each wave: with shorter spacing between waves in the early years, the reduction in uncertainty due to each wave is likely to be smaller, because less change is expected to occur. The appropriate design needs to balance the smaller uncertainty reduction associated with shorter between-wave spacing in the early years against the longer utility of the information obtained.

Motivated by the design issues in the GWVHS, we have conducted some preliminary research that indicates the variable-frequency design can be beneficial, especially if the uncertainty is high prior to the study. Further research in this area would be worthwhile, not only to address the design issues in the GWVHS, but also to address similar public-health-oriented issues.

SURVEY MODALITY AND GEOGRAPHICAL CLUSTERING

With the revolution in communication and information technologies in recent years, there are many modalities to be considered for conducting a survey study, including face-to-face interview, telephone interview, mail administration, and combinations of those modalities. Computer-assisted interview is commonly used in the face-to-face and telephone interviews. Audio-assisted interview is sometimes used in the face-to-face modality for sensitive topics, for the respondents to be interviewed privately without direct interface with the interviewer. Internet-based interview via e-mail or the World Wide Web is becoming a possibility. The choice of survey modality can affect both the response rate and the quality of the data (see, e.g., Groves, 1989, Chapter 11; McHorney et al., 1994; Weinberger et al., 1994, 1996; and Wu et al., 1997).

Face-to-face interview is usually considered to be the most reliable modality. This modality allows for the use of auxiliary material such as printed response categories and prompts that are difficult to use under the telephone modality. The face-to-face contact between the interviewer and the respondent also facilitates the retrieval of written or printed documents such as insurance benefit brochures. This modality also allows for direct examination of the respondents, such as the taking of physical measurements. There is some evidence that this modality might result in more socially desirable responses, such as underreporting of stigmatized behavior. Most importantly, this is usually the most costly modality, because it requires the physical co-location of the interviewer and the respondent at the same time.

Telephone interview is less costly than face-to-face interview. It is limited by the access to the respondent by telephone; thus, it is difficult to utilize for respondents who do not have direct access to a telephone. Mail survey is usually the least costly, but usually results in a higher nonresponse rate than the other modalities.

An important design implication of the face-to-face modality is that it is usually economical to implement this modality using a geographically clustered sample, to reduce the travel cost for the interviewers. More specifically, the sample is usually drawn in multiple stages; a sample of geographical locales is drawn first, usually using the probability-proportional-to-size (PPS) design. A sample of individuals is then drawn from each sampled locale.

There are several other design features in the GWVHS that might also indicate the need for a geographically clustered sample. The case-control portion of the GWVHS will also require geographical clustering, to facilitate the identification of non-GW veteran cases with medically unexplained physical symptoms. In order to maximize the comparability between the GWVHS sample and the non-GW veterans sample with medically unexplained physical symptoms, it would be desirable to take a portion of the GWVHS sample from the same geographical locales where the non-GW veteran cases with medically unexplained physical symptoms are sampled. In addition, if physical examinations will be conducted on a subsample of GWVHS participants, it would be worth considering making the subsample geographically clustered, to facilitate the face-to-face contact for the physical examination. Finally, some of the GWVHS analyses might require the use of contextual data, such as the supply of health care in the geographical locale in which the respondent resides. While some contextual data can be obtained with little effort for all geographical locales, some detailed contextual data would require direct data collection from the agencies in the geographical locales being studied. Geographical clustering of the GWVHS sample will also reduce the cost of this type of data collection.

The clustering of the sample usually reduces the statistical information available in the sample, due to the intracluster correlation (ICC): the respondents from the same geographical locale are usually more similar to each other than the respondents from different locales. Due to the ICC, the statistical information obtained from each respondent usually exhibits diminishing return with the sample size in the same locale: the second respondent from a locale contributes less information than the first respondent, the third respondent contributes less than the second, and so on. In the extreme case of perfect ICC (all respondents from the same locale have identical attributes), the first respondent conveys all information available; subsequent respondents from the same locale contribute no additional information. Therefore the design of a geographically clustered sample needs to balance between the cost savings due to clustering and the diminishing return in the statistical information (see, e.g., Kish, 1965, Chapter 5).

It is challenging to maintain the geographically clustered design in a panel study: the level of clustering usually dissipates gradually over time, due to migration.[15] Some participants will migrate out of the original locales they were recruited from. Some of the migrants might move to a new geographical locale that is part of the sampled locales, or near a sampled locale; thus, they can be followed easily using the face-to-face modality. Some of the migrants might move to a new geographical locale that is not near any sampled locales. It will be costly to follow those migrants using the face-to-face modality.

To the extent that face-to-face contact is important, either for the entire GWVHS sample or for a subsample, the dissipation of geographical clustering might make it prohibitively costly to follow the sample over time, thus making it necessary to draw a new panel.

NONRESPONSE, ATTRITION, TRACKING, AND TRACING

Nonresponse is a common challenge to the implementation of most survey studies, both for cross-sectional and longitudinal studies (see, e.g., Bailar, 1989; Brick and Kalton, 1996; Groves, 1989, 1998; Kalton, 1986; Kish, 1965, Chapter 13; Laird, 1988; Lepkowski, 1989; and Lessler and Kalsbeek, 1992). For longitudinal studies, nonresponse can occur both at the baseline and in subsequent follow-up surveys. We begin our discussions with baseline nonresponse, followed by attrition (follow-up nonresponse), and techniques commonly used to reduce nonresponse.

The GWVHS is likely to be especially vulnerable to baseline nonresponse. As this report is being written, 8 years have elapsed since the Gulf War. While DoD maintains the locating information on records for all veterans, including the GW veterans, the accuracy of this information is likely to deteriorate over time. We anticipate, therefore, that a substantial tracking and tracing effort will be necessary to recruit a representative sample of GW veterans, and a substantial level of baseline nonresponse will still occur despite this effort.

A substantial level of baseline nonresponse raises concerns about nonresponse bias. The nonrespondents are likely different from the respondents, thus the data obtained from the respondents might be biased and fail to represent the target population appropriately. To a limited extent, baseline nonresponse bias can be mitigated using nonresponse weighting and poststratification weighting, (multiple) imputation, pattern mixture modeling, and selectivity modeling (see, e.g., Brick and Kalton, 1996; Copas and Farewell, 1998; Heckman, 1979; Little, 1993, 1994; Little and Rubin, 1987; Rubin, 1987, 1996; and Schafer, 1997). Those techniques usually rely on strong assumptions that are difficult to verify empirically. Nonresponse weighting and imputation usually relies on the assumption that the nonrespondents are missing at random (MAR) after controlling for characteristics observed for both the respondents and nonrespondents. Pattern mixture models provide a way to assess the sensitivity of the analysis

[15]Given the limitations in the DoD data, it is likely that migration already took place for some individuals sampled for the first wave.

results to informative missingness that violate the MAR assumption. Selectivity models usually rely on assumptions about the joint distributions about the nonresponse mechanism and the outcome data to assess the informative missingness. Those assumptions should be viewed as (at best) approximations to the reality. If the nonresponse rate is low, those techniques might help reduce the nonresponse bias. If the nonresponse rate is high, the potential for nonresponse bias is likely to remain even after applying those techniques.

As an example of those techniques, nonresponse weighting is based on comparisons between respondents and nonrespondents, using characteristics available for both groups, such as the administrative information available from DoD. A nonresponse weighting model[16] is then used to estimate the response rate for each category of subjects; the reciprocal for the estimated response rate is used as the nonresponse weight. The subjects with low estimated response rates are given larger nonresponse weights; those with high estimated response rates are given smaller nonresponse weights. The nonresponse weights therefore adjust the distribution in the observed sample so as to compensate for the distortion resulting from the nonresponse. Similarly, poststratification weighting compares the respondents with the target population using characteristics available for both groups, to adjust for distortion that might have occurred in the observed sample.

Among respondents to the baseline survey, some might fail to respond to some or all of the subsequent follow-up surveys. Many panel surveys accomplish very high follow-up rates (95% or more) from wave to wave. However, nonresponse usually accumulates across waves, thus the overall response rate might be substantially lower after many waves. Attrition becomes a more severe problem.

Like baseline nonresponse, nonresponse during the follow-up waves is also likely to result in bias in the observed data. In particular, the cross-sectional representativeness of the cohort is likely to deteriorate over time, as the baseline nonresponse is compounded with attrition. This is one of the main reasons many panel studies use the repeated panel design to regain the cross-sectional representativeness. To a limited extent, wave nonresponse bias can also be mitigated using the techniques discussed earlier for baseline nonresponse (see, e.g., Copas and Farewell, 1998; Diggle, 1989; Diggle and Kenward, 1994; Diggle et al., 1994; Heckman and Robb, 1989; Kalton, 1986; Kyriazidou, 1997; Laird, 1988; and Lepkowski, 1989).

Many panel surveys restrict the follow-up to the respondents to the previous wave, thus the nonrespondents to the baseline survey are eliminated from all follow-ups, the nonrespondents to the first follow-up survey are eliminated from all

---

[16]The nonresponse weighting model is either a simple cross-tabulation (an ANOVA model) that stratifies the respondents and nonrespondents according to a few available characteristics such as gender, race/ethnicity, and age bracket, or a logistic regression model that predicts the response status (being a respondent vs. a nonrespondent) using a larger number of available characteristics.

subsequent follow-ups, and so on. This practice is usually based on the expectation that the nonrespondents are unlikely to convert into respondents in future waves.[17] While this expectation is not unreasonable, this practice usually results in a substantial accumulation of nonresponse across waves. In order to reduce the level of cumulative attrition, we believe it would be appropriate to make efforts to survey the nonrespondents to earlier waves (including the baseline nonrespondents), unless they have given explicit instructions not to be contacted again.

There are a number of procedures commonly used in panel studies to reduce nonresponse, namely, tracing and tracking (see, e.g., Burgess, 1989).[18] Given the anticipated difficulties in recruiting the GW veterans, those procedures are important to help strengthen the quality of the GWVHS data.

Tracing is in essence looking for a missing person using available information. Prospective tracing is necessary at the baseline to locate the participants who cannot be located using the record data from DoD information. Retrospective tracing is necessary for participants who were surveyed in an earlier wave but could not be located at a subsequent follow-up wave.

A variety of public information sources are usually used for tracing, such as telephone directories, credit records, property records, court records, mortality records, (to identify deceased sampled subjects), and so on. It is important in the tracing procedure to verify the identity of the subjects located, to avoid false identification.

Customized tracing procedures can also be utilized, such as visiting the subject's prior residences, neighbors, and known and possible associates. Those procedures are labor intensive, and thus are likely to be too costly for the national scope of the GWVHS sample. It is conceivable that those procedures can be utilized for participants clustered in a limited number of geographical areas if a (partially) clustered design is used for the GWVHS.

While tracing is used to locating the missing subjects, tracking is used to maintain contact with subjects already located. During the baseline survey, the interviewer collects contact information (both primary and secondary) from the participants, to help locate the participants for subsequent follow-up surveys. The contact information is usually updated during each follow-up survey. For studies with infrequent survey frequencies, additional tracking procedures are usually deployed to maintain contact with the participants between waves. This includes sending postcards, birthday cards, and newsletters to the participants at regular intervals, requesting postal notification of change of address, and re-

---

[17] Another reason for the "monotonic follow-up" might be the anticipation that the cases with incomplete data will be eliminated from the analysis. While this might be true for the way longitudinal data was analyzed, the analysis techniques developed in recent years, such as multilevel modeling, do not require all respondents to be observed at the same time points; thus the cases missing some waves can be used in the analysis under appropriate assumptions.

[18] We distinguish tracking as following those with whom we have active contact, and tracing as locating those with whom we do not have active contact.

questing the participants to submit change-of-address information to the study (an incentive is usually offered to encourage the participants to provide this information). In addition to updating the contact information, some of those procedures (e.g., birthday cards and newsletters) might also enhance the goodwill among the participants, so as to facilitate their cooperation at subsequent follow-up surveys.

Given the long lag between waves for GWVHS, additional procedures can be utilized to help maintain the contact information. One possibility is to conduct "light-duty" tracing procedures (such as retrieving easy-to-access public records) on the participants regularly. Burgess (1989) recommended that "If the intersurvey period is five years, . . . it may be more cost-effective to trace a person five times over five years than once after five years."

Another possible procedure is to make brief telephone contacts with participants between waves, to greet the participants (hopefully to enhance the goodwill), and to request updates on contact information. It is conceivable that more intensive tracing procedures can also be used between waves to maintain the contact information. Those procedures are more costly, therefore it might be appropriate to restrict them to participants anticipated to be more difficult to follow, such as those who were difficult to locate during an earlier wave.

MEASUREMENT ERROR

Empirical data are almost always subject to measurement error. Survey data are no exception. Some types of measurement errors are general, and apply to both cross-sectional and panel surveys. Some types of measurement errors are specific to panel surveys. We discuss both types of measurement errors, and remedies that can help mitigate problems resulting from those measurement errors (see, e.g., Bailar, 1989; Groves, 1989; Groves and Couper, 1998; Kish, 1965, Chapter 13; and Lessler and Kalsbeek, 1992).

General Measurement Error Issues

Part of the measurement error can be attributable to the respondent. The respondent might intentionally provide an inaccurate response to a survey question. For example, the respondent might intentionally provide a socially desirable response, or refuse to report a stigmatized condition. The inaccurate response might also be given unintentionally, because the respondent does not have the necessary information, or does not want to make the effort to compile the necessary information.

Part of the measurement error can be attributable to the interviewer—this is applicable to face-to-face and telephone interviews delivered by an interviewer, it is not applicable to self-administered mail surveys. For example, the interviewer might not accurately follow the branching logic to deliver the appropriate survey questions to the respondent; might not convey the survey question clearly to the respondent, might not guide and motivate the respondent to compile and process the information necessary to provide accurate responses, might record the respondent's responses erroneously, or might not be alert in identifying inconsistencies in the respondent's responses and request the respondent to confirm the responses. In the worst scenario, the data might be forged in part or in its entirety by the interviewer.

Part of the measurement error can be attributable to the survey instrument. For example, the branching logic in the survey instrument might be inappropriate, leading the respondent to miss applicable questions; the survey questions might not be organized in a user-friendly sequence to make it easy for the respondent to compile and process the information accurately; the wording of the survey question might not be cognitively clear, resulting in confusion and misinterpretation by the respondent; the response categories might not be defined clearly (mutually exclusive and exhaustive) for the respondent to classify his or her status according to the given categories.

Finally, part of the measurement error might be attributable to data processing subsequent to the interview, such as data entry errors, coding errors, secondary errors introduced in data editing, errors in matching records, and so on.

The nature of measurement error can usually be classified into systematic error and random error. The quality of survey responses is usually characterized using validity and reliability: validity measures the level of systematic error, reliability measures the level of random error. We use the term "accuracy" below to refer to the combination of validity and reliability.

The level of measurement error can be evaluated using a number of techniques, such as test–retest (to assess reliability), comparisons with alternative data sources such as records data (to assess validity), and so on.

Systematic error occurs when similar measurement error persists across multiple waves of surveys, and/or when similar measurement error occurs across respondents. For example, the respondents might overreport outpatient medical visits systematically. Systematic error usually leads to bias in estimated population parameters such as the prevalence for a disease condition or the average level of outpatient service use.

Random error usually varies over time across multiple waves of surveys, and/or varies across respondents. For estimating aggregate population parameters such as the disease prevalence or average service use, random error usually results in reduced power and precision, but does not result in bias. However, random error might result in the overestimation for individual level gross changes. There are many techniques and procedures that can be used to mitigate measurement error in the survey data. We describe several below.

Many of the general sources of measurement errors can be mitigated with computer-assisted survey techniques. For example, the computerized survey instrument usually incorporates built-in branching logic, thus avoiding interviewer mistakes in following the branching logic. It is of course still crucial that the branching logic be designed accurately and programmed accurately. Data entry errors are essentially eliminated in computerized surveys, to the extent the interviewer records the respondent's responses accurately.

Thorough interviewer training and monitoring is essential to mitigate measurement error. In addition, the match between the interviewer and the respondent can help improve the rapport for the interview, such as the match in race and ethnicity or the use of HIV-positive interviewers in surveys of HIV-positive respondents.

In-depth cognitive testing of the survey instrument can be used to identify ambiguities in the wording of the survey questions and response categories. The results of the cognitive testing can be used to revise the instrument, improve the clarity, and reduce the measurement error. Similar laboratory-based testing of other design features about survey questions and instruments, such as the sequential order of survey questions in an instrument, can also help address potential measurement error issues.

An alternative to laboratory-based testing prior to the deployment of the survey is to include substudies in the survey study to assess important measurement error issues. As an example, the RAND Health Insurance Experiment (Newhouse et al., 1993) included a substudy on the frequency of health reports, randomizing the respondents to various levels of reporting frequency, to assess the potential that the health report might prompt the respondents to seek medical care.

Some sources of measurement error can be mitigated with the appropriate choice of survey modality. For example, audio-assisted interview can be incorporated into the face-to-face modality for sensitive and stigmatized topics to reduce the respondent's concern about providing socially undesirable responses to the interviewer. Sometime a randomized response design (Horvitz, 1967) is used, in which the respondent's response is randomized to help alleviate his concerns.

For survey questions that inquire about the respondent's past experience or future anticipation, the level of measurement error is determined by the reference period (see below). Therefore it is important to make an effort to choose an appropriate reference period to reduce the measurement error.

Reference Period

The nature of the measurement error for a specific attribute usually depends on the time frame for the trait being measured. Some traits are usually time-invariant, such as birth year, gender, race, and ethnicity. Time frame is usually irrelevant for those traits. Some traits vary over time, therefore the specific measure needs to take the time frame into consideration. Some measures are specific to the current status and thus can be viewed as snapshots, such as current health status (excellent, good, fair, poor), current marital status, and current employment status. Some measures inquire about events that occurred during the specified time interval (the *reference period*), such as the number of outpatient medical visits during the last 6 weeks. Some measures inquire about the accumulation or the central tendency of a time-variant trait over a reference period, such as the household income during the calendar year 1998 (accumulated over time), the number of cigarettes smoked each day (on the average) over the last 6 weeks, and the level of satisfaction with the primary care provider during the last 6 weeks (presumably the central tendency over this time interval). Most survey questions are retrospective and inquire about reference periods that occurred in the past. Some might be prospective and inquire about the respondent's anticipation for the future. The reference period might be a fixed time interval determined by the calendar (such as the calendar year 1998), a fixed time interval defined relative to the time of the interview (the last 6 weeks, the next 2 weeks), a time interval defined relative to an easily recognized milestone (since the last interview, since the most recent discharge from a hospital, until the anticipated surgery). Note that the duration for the milestone-based reference periods might vary from respondent to respondent; it might even be unknown (for prospective milestones). The analysis needs to take those variations into account—there is more chance for events to occur in a longer reference period. A special type of milestone-based reference period is the lifetime experience (since the respondent's birth), or lifetime anticipation (till the respondent's death).[19]

The nature of the reference period for a specific survey measure has important implications on the measurement error (see, e.g., Bailar, 1989; Neter and Waksberg, 1964). The respondent might misclassify the time for specific events relative to the reference period, resulting in telescoping (inclusion of events that occurred outside the reference period) and omission of events that occurred inside the reference period. Omission can also occur irrespective of the reference period: the reason for the omission might be the respondent's failure to recognize or report a specific event, rather than the respondent's failure to classify the time for the event accurately. "Fabrication" of nonexistent events can also occur irrespective of the reference period.

The accuracy of the survey response usually decreases with the length of the reference period: both telescoping and omission are more likely to occur when the respondent is required to recall or anticipate events distant in the past or the future. Exceptions to this general rule might occur if the longer reference period is easier for the respondent to recognize. For example, it might be easier for the respondent to report taxable income for the calendar year 1998 (the available information is likely organized by calendar year) rather than for the last 6 months (the respondent might have difficulties determining whether specific payments were received within or prior to the last 6 months). The presence of milestones might also help the respondent to respond accurately, even though the reference period might be longer than an alternative shorter fixed reference period.

---

[19]The analysis of lifetime experience data needs to take the duration into consideration. For example, the lifetime prevalence for a specific disease condition is likely lower for respondents in their 30s than respondents in their 40s, because the latter group had more time to develop the condition.

The appropriate reference period to be used in a survey question depends on the trait being inquired. For major events such as hospitalization (to be more specific, discharge from a hospital), the accuracy of the respondent's recall usually remains high even for long time periods of 6 months or a year. For less "impressive" events such as outpatient visits, the accuracy might deteriorate substantially beyond a few weeks.

In addition to the accuracy of the survey measure, the choice of the appropriate reference period should also take into consideration the sampling variation associated with the time frame. In the absence of measurement error, the statistical information in the survey measure increases with the length of the reference period. In a sense, the effective sample size should be measured in terms of person-time. The longer reference period allows more events to be accumulated, thus provides more information. For example, it is conceivable that we can obtain nearly perfect reporting of hospital discharges during the last 7 days. However, very few respondents experienced hospital discharges during such a short reference period, therefore the precision of the estimated rate of hospitalization will be poor due to the high sampling error.

The rate of hospitalization needs to be defined relative to time, such as the number of hospital discharges per thousand person-years. A sample of a thousand individuals asked about a 7-day reference period only contributes about 20 person-years' worth of data. The same sample asked about a 12-month reference period will contribute a thousand person-years' worth of data. The latter design might be preferable even if the measurement error might be larger with the 12-month reference period. The ultimate choice of the time interval needs to be based on the trade-off between the reduction in the sampling error and the increase in measurement error due to the use of the longer time interval.

Most data items to be used in the GWVHS will likely be standard, with known properties in the quality of the recall measures. If some recall data elements are new, or if there are concerns about the recall properties among the GW veterans for some existing data elements, one might consider a substudy to use different recall periods for randomly partitioned subsamples, say, inquire about 3 months for a random subsample, and 6 months for others. The consistency between the two versions of the survey instrument can then be tested using the two subsamples.

Measurement Error Issues Specific to Panel Surveys

One of the advantages in using panel surveys is that the previous interview and/or events reported during the previous interview can be used as milestones to bound the reference period, to help improve the accuracy of the respondent's recall. This advantage is applicable when the reference period coincides with the lag between successive waves of surveys. However, this advantage is unlikely to be applicable to the GWVHS. The anticipated lag time between waves for GWVHS is much longer than the reference periods appropriate for most health outcome measures, therefore it is unlikely that prior interviews can be used as milestones to bound the reference period for subsequent interviews.

A unique measurement error issue for panel surveys is panel conditioning: the observed responses might be affected by the participation in the panel, thus compromising the validity of the data obtained in later waves (see, e.g., Bailar, 1975, 1989; Cantor, 1989; Corder and Horvitz, 1989; Holt, 1989; Presser, 1989; Silberstein and Jacobs, 1989; and Waterton and Lievesley, 1989).

There are a number of possible interpretations for the panel conditioning. First, the participation in the panel might affect the respondents' actual behavior. For example, the survey might serve as a prompt for the participants to attend to their health care needs. Under this scenario, the survey responses in the subsequent waves might reflect the actual behavior and its consequences, but the behavior might not be representative of what would take place in the general population in the absence of the earlier survey. The potential for the participation effect is especially important if physical examination is conducted on a subsample of the GWVHS participants: the physical examination might reveal a health condition that requires medical care, thus having an impact on the health status for the participants in this subsample. The impact might be both short term and long term: the medical care received might affect the trajectory of the health status.

Second, the participants might learn from earlier waves that certain "trigger" items would lead to additional items; they might avoid the burden by responding to the "trigger" items negatively in future waves to avoid the additional items. Under this scenario, the survey responses in the subsequent waves would be biased towards underreporting of the "trigger" conditions.

Third, the participants might learn from earlier waves what is the information required for the survey; thus, they are more capable of compiling and processing the information required to provide accurate responses to the survey questions. Under this scenario, the panel conditioning will reduce the measurement error.

The presence of panel conditioning is easy to detect under the rotating panel design. For each wave of survey, we have respondents at various levels of "seniority" on the panel: some are new, some have had some experience on the panel in earlier waves, some have completed their tenure on the panel and are ready to retire from the panel. We can therefore compare the responses given by respondents at various levels of "seniority" on the panel to assess the presence of panel conditioning. (It is important, though, to control for attrition in those comparisons.) If panel conditioning is judged to be important, the rotating panel design should be considered to make it easy to address panel conditioning.

It is much more difficult to assess panel conditioning with either the permanent panel design or repeated panel surveys without temporal overlap. The comparison across waves cannot be used to assess panel conditioning because it is confounded with true changes over time. It is conceivable that some comparisons with records data can be made, maybe for a subsample, to assess the measurement error due to panel conditioning. This will not address the impact of panel conditioning on actual behavior.

It is possible to assess panel conditioning by using a substudy that varies the follow-up frequency. For example, we can take a random subsample and interview them at a higher frequency, say, annually, to compare their responses to the responses in the rest of the sample. This might be too costly to be worthwhile. Of course we also obtain more data in the subsample; thus, we might be able to reduce the overall sample size.

The lag between waves is anticipated to be fairly long for the GWVHS. Therefore it is plausible that panel conditioning is unlikely to happen, except for the long-term impact of the physical examination. Therefore, we should focus the assessment of panel conditioning on the impact of the physical examination, and place a low priority on the other components of panel conditioning. More specifically, if we do not detect a long-term impact due to the physical examination, it would be reasonable to assume the absence of other components of panel conditioning. If we do detect a long-term impact due to the examination, we might need to consider either rotating the panel or switching to a new panel.

If physical examination is to be conducted in the GWVHS, it should be designed as a randomized substudy for GWVHS, with a random subsample assigned to receive the examination. It will then be easy to assess the long-term impact of the examination, by comparing the health status in the subsample versus the rest of the sample.

REFERENCES

Armitage, P. 1960. Sequential Medical Trials. Springfield, Illinois: Thomas.

Bailar, B.A. 1989. Information Needs, Surveys, and Measurement Errors. In: *Panel Surveys*, (Eds. D. Kasprzyk, G. Duncan, G. Kalton, and M.P. Singh). New York: John Wiley. Pp. 1–25.

Bailar, B.A. 1975. The Effects of Rotation Group Bias on Estimates from Panel Surveys. *Journal of the American Statistical Association* 70(349):23–30.

Brick, J.M., and Kalton, G. 1996. Handling Missing Data in Survey Research. *Statistical Methods in Medical Research*. 5:215–238.

Burgess, R.D. 1989. Major Issues and Implications of Treating Survey Respondents. In: *Panel Surveys* (Eds. D. Kasprzyk, G. Duncan, G. Kalton, and M.P. Singh). New York: John Wiley. Pp. 52–75.

Cantor, D. 1989. Substantive Implications of Longitudinal Design Features: The National Crime Survey as a Case Study. In: *Panel Surveys* (Eds. D. Kasprzyk, G. Duncan, G. Kalton, and M.P. Singh). New York: John Wiley. Pp. 25–51.

Coad, D.S., and Rosenberger, W.F. 1999. A Comparison of the Randomized Play-the-Winner Rule and the Triangular Test for Clinical Trials with Binary Responses. *Statistics in Medicine* 18:761–769.

Copas, A.J., and Farewell, V.T. 1998. Dealing with Non-Ignorable Non-Response by Using an "Enthusiasm-to-Respond" Variable. *Journal of the Royal Statistical Society, Series A* 161(3):385–396.

Corder, L.S., and Horvitz, D.G. 1989. Panel Effects in the National Medical Care Utilization and Expenditure Survey. In: *Panel Surveys* (Eds. D. Kasprzyk, G. Duncan, G. Kalton, and M.P. Singh). New York: John Wiley. Pp. 304–319.

Day, N.E. 1969. Two-Stage Designs for Clinical Trials. *Biometrics* 25:111–118.

Diehr, P., Martin, D.C., Koepsell, T., Cheadlee, A., et al. 1995. Optimal Survey Design for Community Intervention Evaluations: Cohort or Cross-Sectional? *Journal of Clinical Epidemiology* 48(12):1461–1472.

Diggle, P.J. 1989. Testing for Random Dropouts in Repeated Measurement Data. *Biometrics* 45:1255–1258.

Diggle, P.J., and Kenward, M.G. 1994. Informative Drop-Out in Longitudinal Data Analysis. *Applied Statistics* 43(1):49–93.

Diggle, P.J., Liang, K-Y., and Zeger, S.L. 1994. *Analysis of Longitudinal Data*. Oxford: Clarendon Press.

Duncan, G.J., and Kalton, G. 1987. Issues of Design and Analysis of Surveys across Time. *International Statistical Review* 55(1):97–117.

Duncan, G.J., Juster, F.T., and Morgan, J.N. 1984. The Role of Panel Studies in a World of Scarce Research Resources. In: *The Collection and Analysis of Economic and Behavior Data* (Eds. S. Sudman and M.A. Spaeth). Champaign, Ill.: Bureau of Economic and Business Research & Survey Research Laboratory. Pp. 94–129.

Gail, M.H., Mark, S.D., Carroll, R.J., Green, S.B., and Pee, D. 1996. On Design Considerations and Randomization-Based Inference for Community Intervention Trials. *Statistics in Medicine* 15:1069–1092.

Groves, R.M. 1989. *Survey Errors and Survey Costs*. New York: John Wiley.

Groves, R.M., and Couper, M.P. 1998. *Nonresponse in Household Interview Surveys*. New York: John Wiley.

Heckman, J.J. 1979. Sample Selection Bias as a Specification Error. *Econometrica* 47:153–161.

Heckman, J.J., and Robb, R. 1989. The Value of Longitudinal Data for Solving the Problem of Selection Bias in Evaluating the Impact of Treatments on Outcomes. In: *Panel Surveys* (Eds. D. Kasprzyk, G. Duncan, G. Kalton, and M.P. Singh). New York: John Wiley. Pp. 512–539.

Hedeker, D., Gibbons, R.D., and Waternaux, C. 1999. Sample Size Estimation for Longitudinal Designs with Attrition: Comparing Time-Related Contrasts Between Two Groups. *Journal of Educational and Behavioral Statistics*, in press.

Hirano, K., Imbens, G.W., Ridder, G., and Rubin, D.B. 1998. Combining Panel Data Sets with Attrition and Refreshment Samples, NBER Technical Working Paper No. 230. Pp. 1–37.

Holt, D. 1989. Panel Conditioning; Discussion. In: *Panel Surveys* (Eds. D. Kasprzyk, G. Duncan, G. Kalton, and M.P. Singh). New York: John Wiley. Pp. 340–347.

Horvitz, D.G., Shah, B.V., and Simmons, W.R. 1967 The Unrelated Question Randomized Response Model. American Statistical Association: Proceedings of the Social Statistics Section, Pp. 67–72.

Hsiao, C. 1986. *Analysis of Panel Data*. New York: Cambridge University Press.

Hsiao, C. 1985. Benefits and Limitations of Panel Data. *Economic Reviews* 4(1):121–174.

Jabine, T.B., King, K.E., Petroni, R.J. 1990. Survey of Income and Program Participation Quality Profile. Bureau of the Census, U.S. Department of Commerce, Washington D.C.

Kalton, G. 1986. Handling Wave Nonresponse in Panel Surveys. *Journal of Official Statistics* 2(3):303–314.

Kalton, G., and Citro, C.F. 1993. Panel Surveys: Adding the Fourth Dimension. *Survey Methodology* 19(2):205–215.

Kish, L. 1965. *Survey Sampling*. New York: John Wiley.

Kyriazidou, E. 1997. Estimation of a Panel Data Sample Selection Model. *Econometrica* 65:1335–1364.

Lai, T.L., Levin, B., Robbins, H., and Siegmund, D. 1980. Sequential Medical Trials. *Proceedings of the National Academy of Sciences USA* 77(6):3135–3138.

Laird, N.M. 1988. Missing Data in Longitudinal Studies. *Statistics in Medicine* 7:305–315.

Lepkowski, J.M. 1989. Treatment of Wave Nonresponse in Panel Surveys. In: *Panel Surveys* (Eds. D. Kasprzyk, G. Duncan, G. Kalton, and M.P. Singh). New York: John Wiley. Pp. 348–374.

Lessler, J.T., and Kalsbeek, W.D. 1992. Nonsampling Error in Surveys. New York: John Wiley.

Little, R.A. 1993. Pattern-Mixture Models for Multivariate Incomplete Data. *Journal of the American Statistical Association* 88:125–134.

Little, R.A. 1994. A Class of Pattern-Mixture Models for Normal Incomplete Data. *Biometrika* 81(3):471–483.

Little, R.A., and Rubin, D.B. 1987. *Statistical Analysis with Missing Data*. New York: John Wiley.

Maxwell, S.E. 1998. Longitudinal Designs in Randomized Group Comparisons: When Will Intermediate Observations Increase Statistical Power? *Psychological Methods* 3(3):275–290.

McHorney, C.A., Kosinski, M., and Ware, J.E. 1994. Comparisons of the Costs and Quality of Norms for the SF-36 Health Survey Collected by Mail versus Telephone Interview: Results from a National Survey. *Medical Care* 32(6):351–367.

Neter, J., and Waksberg, J. 1964. A Study of Response Errors in Expenditure Data from Household Interviews. *Journal of the American Statistical Association* 59:18–55.

Newhouse, J.P., and the Insurance Experiment Group. 1993. Free for All? Lessons from the RAND Health Insurance Experiment. Cambridge Massachusetts: Harvard University Press.

Overall, J.E., and Doyle, S.R. 1994. Estimating Sample Sizes for Repeated Measurement Designs. *Controlled Clinical Trials* 15:100–123.

Pierce, P. 1997. Physical and Emotional Health of Gulf War Veteran Women. *Aviation, Space, and Environmental Medicine,* P. 68.

Presser, S. 1989. Collection and Design Issues: Discussion. In: *Panel Surveys* (Eds. D. Kasprzyk, G. Duncan, G. Kalton, and M.P. Singh). New York: John Wiley. Pp. 75–79.

Robbins, H. 1974. A Sequential Test for Two Binomial Populations. *Proceedings of the National Academy of Sciences, USA* 71:4435–4436.

Rubin, D.B. 1987. *Multiple Imputation for Nonresponse in Surveys*. New York: John Wiley.

Rubin, D.B. 1996. Multiple Imputation after 18+ Years. *Journal of the American Statistical Association* 91:473–489.

Schafer, J.L. 1997. *Analysis of Incomplete Multivariate Data*. London: Chapman and Hall.

Silberstein, A.R, and Jacobs, C.A. 1989. Symptoms of Repeated Interview Effects in the Consumer Expenditure Interview Survey. In: *Panel Surveys* (Eds. D. Kasprzyk, G. Duncan, G. Kalton, and M.P. Singh). New York: John Wiley. Pp. 289–303.

Stretch, R.H., Bliese, P.D., Marlowe, D.H., Wright, K.M., Knudson, K.H., and Hoover, C.H. 1995. Physical Health Symptomatology of Gulf War-Era Service Personnel from the States of Pennsylvania and Hawaii. *Military Medicine* 160:131–136.

Waterton, J., and Lievesley, D. 1989. Evidence of Conditioning Effects in the British Social Attitudes Panel. In: *Panel Surveys* (Eds. D. Kasprzyk, G. Duncan, G. Kalton, and M.P. Singh). New York: John Wiley. Pp. 319–339.

Wei, L.J., and Durham, S. 1978. The randomized play-the-winner rule in medical trials. *Journal of the American Statistical Association* 73:840–843.

Weinberger, M., Nagle, B., Hanlon, J.T., Samsa, G.P., et al. 1994 Assessing Health-Related Quality of Life in Elderly Outpatients: Telephone versus Face-to-Face Administration. *Journal of the American Geriatrics Society* 42:1295–1299.

Weinberger, M., Oddone, E.Z., Samsa, G.P., and Landsman, P.B. 1996. Are Health-Related Quality-of-Life Measures Affected by the Mode of Administration? *Journal of Clinical Epidemiology* 49(2):135–140.

Weinstein, M.C. 1974. Allocation of Subjects in Medical Experiments. *New England Journal of Medicine* 291:1278–1285.

Whitehead, J. 1997. *The Design and Analysis of Sequential Clinical Trials*, revised 2nd edition. New York: John Wiley.

Wu, A.W., Jacobson, D.L., Berzon, R.A., Revicki, D.A., et al. 1997. The Effect of Mode of Administration on Medical Outcomes Study Health Ratings and EuroQol Scores in AIDS. *Quality of Life Research* 6:3–10.

Zelen, M. 1969. Play the Winner Rule and the Controlled Clinical Trial. *Journal of the American Statistical Association* 64:131–146.